Vacuum Technology, Thin Films, and Sputtering

An Introduction

Vacuum Technology, Thin Films, and Sputtering

An Introduction

R. V. STUART

Koral Labs, Inc.
Minneapolis, Minnesota

1983

ACADEMIC PRESS

A Subsidiary of Harcourt Brace Jovanovich, Publishers

New York   London

Paris   San Diego   San Francisco   São Paulo   Sydney   Tokyo   Toronto

TP 156
V 3
S 78
1983

ACADEMIC PRESS, INC.
111 Fifth Avenue, New York, New York 10003

United Kingdom Edition published by
ACADEMIC PRESS, INC. (LONDON) LTD.
24/28 Oval Road, London NW1 7DX

Library of Congress Cataloging in Publication Data

Stuart, R. V.
    Vacuum technology, thin films, and sputtering.

    Includes index.
    1. Vacuum technology. 2. Thin films. 3. Cathode
sputtering (Plating process) I. Title.
TP156.V3S78 1982   621.5'5   82-13748
ISBN 0-12-674780-6

PRINTED IN THE UNITED STATES OF AMERICA

83 84 85 86   9 8 7 6 5 4 3 2 1

CONTENTS

v

CHAPTER III   VACUUM EVAPORATION

CHAPTER IV   SPUTTERING

CHAPTER V   THIN FILMS

PREFACE

Vacuum technology is advancing and expanding so rapidly that a major difficulty for most companies in this field is finding qualified technicians needed for expansion and as replacements. The only recourse for most companies is to hire capable, though untrained, people and train them in-house. One of the problems in this course of action is that it repeatedly draws on the valuable time of experienced personnel to explain fundamental concepts to a trainee. Even then the trainee is usually at a disadvantage because not more than a glimmer of understanding generally is obtained the first time through, and there is seldom a second time through to provide any more help. Unfortunately there are no available books on vacuum technology written at the introductory level to help fill the void. A rather advanced amount of scientific and technical knowledge is required on the part of the reader to make any real use of presently available books. Even then, a person without experience in this field will generally struggle to find the help actually needed.

This book is written to provide an introduction to vacuum technology in general and the sputtering process in particular. It is written for the novice, whether lab assistant, technician, technologist, engineer, or manager, whose vocation, avocation, and interests necessitate investigating the applications of vacuum technology, thin films, and sputtering. It is expected that those for whom this book is written will have a sufficient technological orientation that their background will enable them to grasp the fundamentals in the manner in which they are presented. It is hoped that experts in the field will find this book helpful in reducing demands on their time for teaching newcomers.

This book mentions by name only a few of the many scientists who have made important contributions in this field. Many additional names could have been justifiably included, but, by the same token, all but one name could reasonably have been omitted. It would have been most improper for a work such as this to neglect to mention G. K. Wehner, who put sputtering on the scientific basis that permitted the development of such an extensive technology. The full contributions by Wehner and other scientists can be found in "Handbook of Thin Film Technology," edited by L. I. Maissel and R. Glang (McGraw-Hill, New York, 1970). This source provides advanced treatment of vacuum technology, vacuum evaporation, and sputtering, including extensive bibliographies. Advanced treatment of thin films may be found in the continuing series of volumes "Physics of Thin Films" (Academic Press, New York) and in "Thin Film Processes," edited by J. L. Vossen and W. Kern (Academic Press, New York, 1978).

The author is grateful to his wife, Evelyn, for her encouragement in writing this book, for valuable suggestions as to content, and for critical evaluation of the manuscript. The employees of Koral Labs, Inc., have been most helpful in reading and evaluating the manuscript. Several companies most generously permitted the discussion of some of their applications in the Specific Applications section. It is our hope that the mention of their names in connection with these applications will help to express our appreciation.

CHAPTER I

EVAPORATION

Introduction

Vacuum technology, thin films, vacuum evaporation, and sputtering deal closely with the atomic domain. We are all somewhat familiar with the ideas of atoms and molecules, but it would not be wise to open our considerations of the subject matter of this book as if we all had a working familiarity with atomic physics. Instead, it is our hope that we shall be able to relate subjects with which we are familiar to the atomic domain in such a way as to find that our knowledge is greater than we may have suspected. We shall use the processes of induction and deduction to gain further knowledge of atoms and molecules, striving always to relate this knowledge to the subject matter of the book. Our intent is to begin with the familiar and lead into areas with which we must become familiar. In keeping with this intent, our first topic of discussion is evaporation. Evaporation, a subject with which we have been familiar since our earliest years, leads directly to discussions of the states of matter. This leads to discussions of atoms and molecules and thence to consideration of vacuums and vacuum technology. A parallel line of discussion leads from evaporation to thin films.

We begin our discussion of evaporation by reminding ourselves that water is commonly encountered in each of the three states of matter: solid (ice), liquid (water), and gaseous (water vapor). We know that heating ice (solid) causes it to turn to water (liquid) and that further heating causes it to turn to water vapor (gaseous). Most other materials are usually thought of as existing in only one of the three states of matter (solid, liquid, or gaseous). Actually, many materials can exist in any of these three states, depending on the temperature. Many materials that we think of as solids can be heated enough to melt (become liquid). Many materials that we think of as liquids can be heated enough to evaporate (become gaseous). We may find it easier to accept this if we first consider evaporating materials that we have encountered or have heard of in both the liquid and gaseous states. Most of us are aware of the existence of a number of materials in both the liquid and the gaseous forms: water and water vapor, gasoline and gasoline fumes, liquid air and air, liquid nitrogen and nitrogen, liquid oxygen and oxygen, liquid hydrogen and hydrogen. We are aware that the difference in state (liquid or gaseous) is dependent on temperature. We know that in order to change the material from the liquid state to the gaseous state, the material must be heated. Thinking of it in this way, we realize that we are familiar with the idea of evaporating a number of materials. It becomes less difficult to believe that, if cooled, these liquids become solids and that these materials therefore can exist in all three states of matter: solid, liquid, and gaseous. Many of us, in fact, are already aware that this is the case.

Now, considering materials that we commonly think of as solids, we can think of cases in which some of these materials exist also in the liquid state. Most of us have heard of, read of, and seen pictures of molten rock (lava) and molten steel. Many of us have seen a metal (solder) melt when heated. When we are reminded that these things are already within our realm of knowledge, we can quite readily accept that many materials can be melted at suitable temperatures. These (suitable) temperatures are quite low for materials that we usually think of as gases, near room temperature for materials that we usually think of as liquids, and quite high for materials that we usually think of as solids.

It now seems to be much less difficult to accept that many materials that we usually think of as solids can be heated enough to melt and can be heated further to evaporate. The characteristic of water that it can exist in any of the three states of matter (solid, liquid, or gaseous) appears to be a characteristic of many materials, and the state in which a given material exists is dependent on its temperature. Just as water can be evaporated and can deposit out of the vapor state (gaseous state) onto the surface of a suitably cool object, many other materials can be evaporated and can deposit out of the vapor state onto the surface of a suitably cool object. Gold, for example, can be evaporated and can deposit out of the vapor state onto the surface of a suitably cool object. A suitably cool temperature for depositing water might be below 0°C, whereas a suitably cool temperature for depositing gold might be anywhere from 1000°C down to near absolute zero, very hot to very cold.

It is probable that many people reading this have been wondering why we have not mentioned mercury. Here is a perfect example of a very-well-known metal that exists in the liquid state at normal temperatures. At lower, but easily attainable temperatures, it exists in the solid state. We have heard of mercury vapor poisoning so we know that mercury exists in the gaseous state. Mercury is a metal that can exist in the solid state at an easily attainable temperature (approximately −40°C), can be converted to the liquid state by heating, and can be converted to the gaseous state by further heating. It becomes very difficult to consider this example and not accept that many other metals should be able to go through these changes in state. Furthermore, if metals can change state from solid to liquid to gaseous, other materials should also be able to do so.

Mercury also helps to lead into the concept of vapor pressure as a function of temperature. We know that mercury vapor poisoning is a potential danger to people who work with mercury or use mercury in their work. The boiling point of mercury is 357°C, but we know that people who have suffered mercury poisoning could not have been exposed to a temperature of 357°C because they would have died of the heat long before poisoning could take effect. It must be concluded that mercury vapor exists at temperatures well below the boiling point. This is not particularly surprising to most of us. We are generally well aware of this situation with respect to water vapor. We at times allow our hands to dry in air so we know that water will evaporate at temperatures below the boiling point. We know that warmer water will evaporate at a faster rate than cooler water. We can easily conclude that water has a vapor pressure that is somewhat proportional to temperature; it has a lower vapor pressure at lower temperatures and a higher vapor pressure at higher temperatures. It is easy to believe that the same is true of mercury. It is not difficult to accept that the same is true of many other materials. One is not greatly astonished to see vapor pressure curves such as those in Figure 1.

At this point we want to remind ourselves that heat, or thermal energy, is the mechanical energy of random motion of atoms. Most of us are aware that atoms and molecules in the gaseous state are in continuous random motion at quite high speeds. Even in the solid state, all atoms and molecules of a material remain in random vibratory motion at all temperatures down to absolute zero. As the temperature increases, this random motion increases, and the average kinetic energy of individual atoms increases. At a certain temperature, characteristic of each given material, this energy is great enough that the attractive forces characteristic of the solid state are overcome, the atoms are no longer retained in fixed positions relative to each other, and the material becomes liquid. The temperature at which this occurs is called the melting point.

In both the solid state and the liquid state, atoms are exchanging energy with each other, sometimes gaining, sometimes losing. It is always possible for an individual surface atom to acquire enough energy through this thermal agitation to escape from the surface. An atom that escapes in this manner from a liquid surface is said to have evaporated. An atom that escapes in this manner from a solid surface is said to have sublimed. The higher the temperature, the greater the probability that evaporation or sublimation can occur. The vapor pressure curves of Figure 1 are, in a sense, indicators of the probability of evaporation or sublimation.

Oxidation

These vapor pressure curves are discussed in a later section. At this time we want to discuss the problem of oxidation, which for many materials occurs when they are heated in air. Water can be

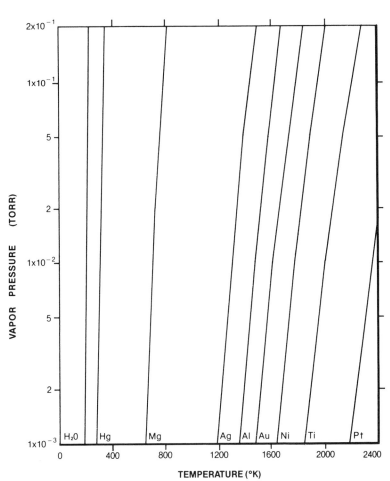

FIGURE 1 Vapor pressure curves.

heated in air without problems because it is already an oxide. Mercury can be heated in air (if you are careful to dispose of the fumes safely) because mercury does not readily oxidize. In fact, if mercury oxide is heated in air, it will decompose into oxygen and mercury. It is common knowledge, however, that most materials (especially metals) will oxidize extensively when heated in air. It is our desire to evaporate the material, not oxidize it. Since oxygen in air is the source of oxidation, it becomes necessary to remove the oxygen. Removing only the oxygen would be a formidable task so we remove the rest of the components of air as well. We find that the removal of all components of air from the area in which we evaporate material provides some side benefits. Furthermore, if we had not removed these other components, we would have found that they would have given side effects as deleterious as oxidation.

Clearly, the air cannot be removed from the area in which the evaporation occurs unless this area is within a sealed chamber. This chamber with the air removed is, of course, now a vacuum chamber. This puts severe constraints on the operation. Everything done in the chamber must be done before the chamber is sealed or must be done by remote control after the chamber is sealed. We must use ingenious methods developed by many people over many years to remove air from the chamber and to keep air from entering the evacuated chamber. These methods are discussed in later chapters. Here, we shall continue to discuss evaporation in air in order to better prepare ourselves to understand evaporation in vacuum.

Evaporative Coating

Just as most of us have a familiarity with evaporation, we have a familiarity with evaporative coating. We may not think of it in these words, but we have seen dew on vegetation and know that this is water vapor that has condensed out of the vapor state in air. We have seen automobile windshields fog up and know that this film of water has condensed out of the vapor state in air. We have seen frost on a windshield and know that this solid film has been deposited out of the vapor state in air. It is (technically) possible to refer to this deposit as a coating, to the object coated as a substrate, and to the process as evaporative coating. Clearly, most of us are familiar with the idea of evaporative coating, i.e., the idea of evaporating material in order to deposit a coating onto a substrate. We at times find the cases with which we are familiar to be aesthetically pleasing, but more often they are a bother and they are never long lasting. Nevertheless, there are many evaporated coatings, both metallic and nonmetallic, that are long lasting and of a beneficial nature, and it is these in which we are interested. As discussed previously, these coatings cannot be deposited in the presence of air.

The coatings in which we are interested must be deposited under vacuum conditions. This is quite different from depositing material out of vapor that is mixed in air, as when water is deposited as dew on roses or frost on windshields. We are familiar with the evaporation of water into the air and the subsequent deposition out of the vapor phase in air onto a flower or a window in the form of dew (liquid) or frost (solid). This process, as we know it, occurs in the presence of air. It is useful to consider this process in air further to see how it differs from evaporation in a vacuum. It is helpful to try to consider this from the point of view of a molecule of water.

Atoms and Molecules

When materials evaporate, they generally leave the liquid state and enter the gaseous state in the form of atoms or molecules. A water molecule emerging from the liquid surface in air sees air molecules everywhere above the surface. We can find some numerical constants that we can use in simplified order of magnitude calculations to get a feel for the sizes, separations, and motions of these molecules. Let us first agree on the simplistic concepts of atoms and molecules that we use in this discussion. We shall think of atoms as being composed of nuclear cores surrounded by shells of electrons. We shall think of nuclear cores as being composed of protons and neutrons, which we shall refer to collectively as nucleons. We shall take the mass of a nucleon as being 1.66×10^{-24} g (gram). We shall take the atomic weights of atoms and the molecular weights of molecules as representing the average number of nucleons in an atom or molecule, and we shall not be at all concerned that these are not whole numbers. We shall take the number of protons in a nucleus as being equal to the number of electrons surrounding the nuclear core. We shall speak of the electrons as being in shells, as being in orbits, or as being in energy states or energy levels. We shall think of molecules as being composed of two or more atoms with electron shells interacting so as to share outermost electrons between or among the atoms.

We now want to apply this simplistic concept to the water molecule. Most of us are aware that a water molecule is composed of two hydrogen atoms and one oxygen atom bound together as symbolized by the chemical formula H_2O. A hydrogen atom is composed of a nucleus made up of a single nucleon (a proton) with a single electron in the peripheral structure of the atom. An oxygen atom is composed of a nucleus made up of 16 nucleons (8 protons and 8 neutrons) with 8 electrons in the peripheral structure. The atomic weight of the hydrogen atom is 1, and the atomic weight of oxygen is 16 so the molecular weight of water is $1 + 1 + 16$ or 18. The molecular weight (18) multiplied by the mass of a nucleon (1.66×10^{-24} g) gives the mass (weight) of a water molecule in grams. At this time we are interested only in relative weights so we shall compare the molecular weights of water and air without converting them to grams.

Most of us know that air is made up of nitrogen, oxygen, argon, water vapor, and some other gases in trace quantities so there is no real molecular weight for air, but we are quite willing to accept an average value. Since air is nearly 80% nitrogen (molecular weight of 28) and about 20% oxygen (molecular weight of 32), then an average molecular weight for air is fairly close to 29. A water molecule emerging from the liquid surface is going to be bouncing around among molecules that are 60% heavier than it is. If it bumps one of them head on or nearly head on, it will be bounced right back to the surface and go back into the liquid phase. Noting the atomic weight (16) of oxygen from the preceding paragraph and the molecular weight (32) of oxygen from this paragraph, we remember that an oxygen molecule is composed of two oxygen atoms. Similarly, a nitrogen molecule is composed of two nitrogen atoms. Inert gasses, such as argon, do not form molecules and exist only in the atomic form.

Diameters of Molecules

In addition to relative weights, we are going to be interested in relative dimensions. A good way to start on this is to first get the actual weight of a water molecule. As indicated previously, this is $18 \times 1.66 \times 10^{-24}$ g or 2.99×10^{-23} g, an incredibly small weight. If one molecule weighs 2.99×10^{-23} g, then a gram of water must contain 3.34×10^{22} molecules of water, an incredibly large number. Any number of references provide the information that the density of water is 1 g/cm³ (gram per cubic centimeter). Combining these two items (a gram of water contains 3.34×10^{22} molecules and a gram of water occupies a volume of 1 cm³) gives the result that 3.34×10^{22} molecules of water in the liquid phase must occupy a volume of 1 cm³. We assume that in both the solid phase and the liquid phase all atoms and molecules are in direct contact with all adjacent atoms and molecules. We picture each atom or molecule as a sphere occupying a little cubicle within the solid or liquid body, as depicted in Figure 2. If we thus assign each molecule of water a little cubicle of space to itself, then the volume of space occupied by a single molecule is 2.99×10^{-23} cm³, and the linear dimensions of this cube are 3.10×10^{-8} cm. We are going to think of the water molecule as a sphere of 3.10×10^{-8} cm diameter.

It is useful to review this method of calculating the size of an atom (or of a molecule). We first find the atomic (molecular) weight M, which is essentially the average number of nucleons in the atom (molecule). We multiply M by the mass of a nucleon (1.66×10^{-24} g) to get the average weight of one atom (molecule). We then

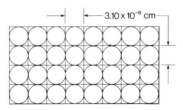

FIGURE 2 Depiction of atoms composing a liquid or a solid as spheres, each occupying a little cubicle of volume.

divide by the density d of the material in solid or liquid form to find the volume occupied by one atom $[(1.66 \times 10^{-24})\ M/d]$. We take the cube root of this and assume that the atom is a sphere of that diameter. It is an implicit assumption in this calculation that each atom or molecule in a solid or liquid is in direct contact with each adjacent atom or molecule. The atomic diameters listed in Table II (see Chapter III) were calculated by this method.

It should be understood that here we are only trying to get a feel for the order of magnitude of relative weights, dimensions, separations, and motions of atoms and molecules. We are going to avoid advanced concepts that may be more descriptive and more accurate but will tend to cloud the primary issue. The simplified ideas used here are relatively easy to understand and give results that are quite adequate. As an example, we can calculate the diameter of an argon atom in the same way that we have calculated the diameter of a water molecule and get a result within 6% of the value given in the "American Institute of Physics Handbook" (AIP Handbook). The atomic weight of argon is 40 so an atom of argon weighs 6.64×10^{-23} g and a gram of argon contains 1.51×10^{22} argon atoms. The density of liquid argon is 1.42 g/cm^3 so a gram of liquid argon occupies a volume of 0.70 cm^3. An argon atom may therefore be assigned a volume of 4.64×10^{-23} cm^3, or a diamter of 3.59×10^{-8} cm. The AIP Handbook gives 3.82×10^{-8} cm as the diameter. One can even see that our calculated value should be on the low side because we have based our calculation on marbles stacked in square array, one directly on another. We know that marbles stacked one directly on another will immediately slip over and stack more tightly. Because of this, the argon atoms, in order to fill up that much volume, have to be a little bigger than our calculated value.

It is quite reasonable to accept the argon atom diameter calculation as being good enough and still question the water molecule diameter calculation. After all, three atoms joined together must have a difficult time forming a spherical shape. The shape of a water molecule with the oxygen atom having two hydrogen atoms sticking out at angles may assume a shape somewhat more like a "Mickey Mouse" head. Nevertheless, the interaction of the electron shells sharing electrons among the three atoms smooths out these bumps to a remarkable extent. Even if this were not the case, we are only trying to get order-of-magnitude values so our result will suffice.

Using the same idea for calculating the average diameter of an air molecule, we remember that the average molecular weight of air is 29. From this we calculate that the weight of an average air molecule is 4.81×10^{-23} g or that a gram of air contains 2.08×10^{22} molecules of air. The density of liquid air is 0.92 g/cm^3 so a gram of liquid air occupies a volume of 1.09 cm^3. An air molecule may therefore be assigned a volume of 5.24×10^{-23} cm^3, or a diameter of 3.74×10^{-8} cm.

It is interesting that these atoms and molecules are all approximately the same diameter even though their weights differ by quite large amounts. We seldom deal with one smaller than 2.5×10^{-8} cm or one larger than 4.0×10^{-8} cm. Basically, the diameter is established by the outer electron shell. The heavier atoms have more electrons and thereore more electron shells, but they also have more highly charged nuclei so that the electron shells are pulled in more strongly. The end result is that the atomic diameter does not increase a great deal as we go to heavier atoms.

Distances between Molecules in the Gaseous State

In the liquid state and in the solid state, atoms and molecules are packed together so that each particle is touching the adjacent particles. This allowed us to calculate atomic dimensions. In the gaseous state, the particles are unbound and touch each other only in chance collisions. To calculate separations between atoms in the gaseous state, we can use the same idea that we used to calculate atomic dimensions in the liquid and solid states. A gram of air contains 2.08×10^{22} molecules of air regardless of whether it is frozen, melted, or evaporated. We can find in reference tables that the density of air at normal temperature and pressure (i.e., in the gaseous state) is 1.29×10^{-3} g/cm^3, from which we calculate that a gram of air in the gaseous state occupies a volume of 775 cm^3. Since 2.08×10^{22} molecules of air in the gaseous state occupy a volume of 775 cm^3, a molecule of air in the gaseous state is surrounded by an otherwise empty volume of 3.73×10^{-20} cm^3, which we think of as a cube of dimension 3.34×10^{-7} cm. If we think of each molecule as being in the center of its cube (on the average), then the average distance between molecules is 3.34×10^{-7} cm. This is a very short distance but not in relative terms; any given molecule can see room for ten more molecules to fit in between it and the next molecule in any direction.

It is useful to review this method of calculating average separations S between atoms in the gaseous state. We multiply the atomic weight M by the mass of a nucleon (1.66×10^{-24} g) to get the average weight of an atom. We then divide by the density of the material in the gaseous state [$d(G)$] to find the volume taken up by one gas atom, $S^3 = (1.66 \times 10^{-24}) M/d(G)$. We take the cube root of this to find the dimension of the cube of volume taken up by one gas atom. We assume that, on the average, each gas atom is at the center of its cube so that the average separation between gaseous atoms is equal to the dimension of the cube: $S^3 = (1.66 \times 10^{-24}) M/ d(G)$ or $S = [(1.66 \times 10^{-24}) M/d(G)]^{1/3}$. We note that the reciprocal of S^3 is equal to n, the number of gas molecules per cubic centimeter.

Going back to the point of view of the water molecule emerging from the liquid surface, we remember that it sees at least room for itself and nine other molecules in every direction. We shall use the ficticious idea that the water surface sees the air above it as layers of thickness 3.34×10^{-7} cm, as depicted in Figure 3. On the average, the nearest "layer" of air molecules is within 3.34×10^{-7} cm of the surface of the water. Since all gaseous molecules are continuously in motion, the water molecule does not really see layers of air molecules, but this is a useful imaginary device for now. The water molecule sees this layer of air molecules divided into squares of 3.34×10^{-7} cm dimension, with each square belonging to an air molecule of 3.74×10^{-8} cm diameter. The area of the square [(3.34×10^{-7} cm)^2, or 1.12×10^{-13} cm^2] is a hundred times as large as the area of the air molecule disk, $\frac{1}{4}\pi(3.74 \times 10^{-8}$ cm)^2 or 1.10×10^{-15} cm^2. This leaves a lot of room for the emerging water molecule to get through after it leaves the water surface. It is not surprising therefore that water can evaporate fairly easily. Actually the water molecule, being of a size comparable to the air molecule,

GASEOUS STATE

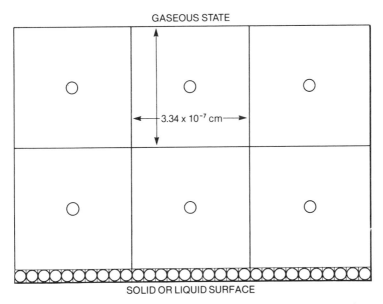

FIGURE 3   Depiction of a solid or liquid surface exposed to air at normal atmospheric pressure.

has a little less escape room than this. The air molecule blocks an area of radius equal to its own radius plus the radius of the water molecule; i.e., $\pi(1.87 \times 10^{-8} + 1.55 \times 10^{-8})^2$, or 3.67×10^{-15} cm^2. Since the ratio of blocked area (3.67×10^{-15} cm^2) to total possibly open area (1.12×10^{-13} cm^2) for the water molecule to get through is 1 to 30, the water molecule risks only 1 chance in 30 of colliding with an air molecule in the first layer. We might guess from this that, on the average, a water molecule will travel through about 15 of our fictional layers of air, a distance of 5.01×10^{-6} cm, before making a collision with an air molecule. This is a very crude and unsophisticated way of looking at this, but the result of much more sophisticated and difficult methods is not much different, 6.81×10^{-6} cm. This "average" distance between collisions is an important concept and is referred to as the mean free path.

It is useful to review this method of calculating the mean free path. We first calculate the average separation S between gas atoms and square this to get S^2, the area of the square within which a typical gas atom is located. We assume that a gas atom can block an area equal to the area of a disk of radius equal to the sum of the radius r of the gas atom and the radius r_t of the atom traveling through the gas, as depicted in Figure 4. This area is $\pi(r + r_t)^2$. The chance that the traveling atom will suffer a collision is simply the ratio R of these two areas, $R = \pi(r + r_t)^2/S^2$. Some atoms may make collisions at the first layer of gas atoms they encounter, but others may pass through $1/R$ layers of gas atoms without making collisions. The rest of the atoms will pass through some intermediate number of layers of gas atoms before making collisions. Averaging all these, we can expect that the average atom will travel through $1/2R$ layers of gas atoms before making a collision. The mean free path (MFP) is thus $S/2R$. Substituting from above for R, we have MFP $= S^3/2\pi (r + r_t)^2$. Remembering that $S^3 = 1/n$, where n is the number of gas atoms per cubic centimeter, we have MFP $= 1/2\pi n(r+r_t)^2$. This is an interesting result in that the mean free path depends only on the number of gas atoms per cubic centimeter, the radius of a gas atom, and the radius of the moving atom. Since the radii of all atomic and molecular particules with which we deal here do not differ much one from another, the factor $(r + r_t)^2$ is a constant for all practical purposes, and the MFP depends only on n, being inversely proportional thereto. If, as an alternative, we

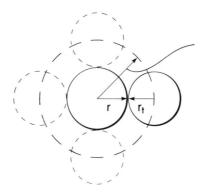

FIGURE 4 Depiction of the idea that a moving atom will be deflected if it approaches within a distance $(r + r_t)$ of another atom.

substitute our earlier expression for S^3, we find MFP $= (8.3 \times 10^{-25}) M/\pi(r + r_t)^2 d(G)$. We shall keep clearly in mind that M, r, and $d(G)$ refer to the gas through which the atomic or molecular particle of radius r_t is moving.

Suppose now that we set out to try to deposit a coating of frost onto a cold pane of glass using a pan of water as the source of vapor. As a matter of convenience we might suspend the pane of glass above the water at a distance of about 10 cm. If a water molecule emerging from the water surface could go in a straight line to the pane of glass, it would, on the average, collide with an air molecule every 6.81×10^{-6} cm (mean free path) and would therefore suffer about 1.5×10^6 collisions on the way. Since each collision would deflect the water molecule, often causing it to re-verse direction, the actual path of a water molecule would be tor-tuous and many, many times as long as the direct-line distance from the pan to the pane. As a matter of fact, before the water molecule has traveled a measurable distance from the water sur-face, it has experienced so many collisions that it has forgotten where it came from and where it is going, and it is much more probable that it will return to the pan then go to the pane.

It begins to appear that the process of evaporative coating in air might have been next to impossible even if we did not have the problem of oxidation. It has been useful to consider this process because we know that, in the case of water, it does indeed occur. Considering this process has enabled us to get some idea of atomic dimensions and perhaps some feel for calculations and numbers in the atomic domain. We are nearly prepared to start considering evaporation in a vacuum, but we first want to see if we can get a feel for the velocities of gas atoms and molecules.

Before proceeding with this, we shall digress a bit (actually a great deal) here in order to insert an order-of-magnitude calcula-tion that we hope will give us some additional understanding of these types of calculations and how they may be related to the macroscopic domain. We shall consider a person who lived a thou-sand or more years ago and whose life span was greater than 20 years. We assume that the average water intake of this person was a liter (1000 cm³) per day. The amount of water that passed through this person's body under these assumptions was therefore in excess of $20 \times 365 \times 1000$ cm^3, or 7.3×10^6 cm^3. We recall that a cubic centimeter of water contains 3.34×10^{22} molecules from which we find that more than 2.44×10^{29} molecules of water passed through this person's body during his lifetime.

We are interested in comparing this quantity with the total number of water molecules in the world. We can make a rough estimate of the total amount of water in the world by estimating the average ocean depth and multiplying by the approximate ocean area. We can find that the deeper regions of the ocean are more than 10,000 m (10^6 cm) deep and the shallower regions about 2000 m deep. If we take 4000 m as an average depth, we cannot be too far off. We frequently encounter references indicating 8000 miles as the diameter of the earth. Converted to more convenient units, this is 1.29×10^9 cm. We can calculate from this that the area of the earth is 5.21×10^{18} cm^2. We are frequently told that the oceans make up ¾ of the total area of the earth so the area of the oceans must be 3.91×10^{18} cm^2. Multiplying by 4000 m (4×10^5 cm) gives 1.56×10^{24} cm^3 as the total amount of water in the world. We could have found in an encyclopedia that the total amount of water in the world is 1.37×10^{24} cm^3, but it is well for us to see that we can often make reasonable estimates when the information we want is not directly available in the desired form. Continuing to use our estimate and the knowledge that a cubic centimeter of water contains 3.34×10^{22} molecules, we see that the total number of water molecules in the world is 5.21×10^{46}. Of these, more than 2.44×10^{29} or a fraction 4.68×10^{-18} passed through the body of our hypothetical person who lived a thousand or more years ago.

We can be sure that these 2.44×10^{29} water molecules made their way ultimately to the oceans either by evaporating and subsequently falling as portions of raindrops or by entering streams and flowing to the ocean. We can find that currents of many kinds exist in the oceans, causing mixing of the ocean waters. These currents range from 10^8 to well over 10^9 cm per year. With our earth diameter being 1.29×10^9 cm or the circumference being 4.0×10^9 cm, the slowest current would result in circumnavigation of the world in 40 years. In 1000 years there could be 25 circuits of the world due to the slowest currents, and there would probably be well over 100. This alone would result in a great deal of mixing of the 2.44×10^{29}

molecules with the total of 5.21×10^{46} molecules, and superimposed on this mixing is the evaporation and rain that has occurred over these years. It is probable that the mixing is quite good and that every volume of water we might consider contains a fraction 4.68×10^{-18} that has passed through the body of our hypothetical person of a thousand or more years ago.

We now consider a small glass of water, specifically a 200 cm³ glass of water. There are $200 \times 3.34 \times 10^{22}$ or 6.68×10^{24} molecules of water in this glass. A fraction of these (4.68×10^{-18}, or a total of 3.13×10^{7} molecules) in all probability had passed through the body of our hypothetical person of a thousand or more years ago. Similar calculations and estimations can be made that indicate that in all probability we are each already recirculating some molecules of water that had once before in our lifetimes been circulated through our bodies. It is the hope here that this somewhat entertaining exercise has impressed on us that our calculations in the atomic domain are indeed related to the macroscopic domain.

Velocities of Gas Molecules

We might, at first, feel that velocities of gas particles would be as outrageous as their sizes (extremely small) or their numbers (extremely large). Actually, velocities of gas particles are of the same order of magnitude as the velocity of sound, a very high speed but one within the range of our experience and ability to measure. We hope that it can be made to seem reasonable that the velocities of air molecules should be at least somewhat similar to the speed of sound. It is common knowledge that sound is generated by something vibrating and passing the vibration on to the air. The vibrating object may be a set of vocal cords, a stereo speaker, a knocked-on door, a dropped watch, etc. We have seen previously that air molecules do not touch each other except by random collisions and are normally widely spaced from each other. We can therefore be sure that the vibrations from the vibrating object are not simultaneously and instantaneously passed on to all air molecules along the path of sound propagation. Instead, these vibrations are transmitted directly only to those gas molecules within the distance of vibration of the vibrating object. We know that such a vibrating object, as, for instance, a knocked-on door, does not have a visible amplitude of vibration (unless it is being knocked on by a highly agitated person). We thus know that only the relatively few gas molecules immediately adjacent to the vibrating object are directly affected by the vibration. These gas molecules are as a group compressed and rarefied as the surface vibrates toward and away from them. By their random motion and random collisions with adjacent air molecules the air molecules directly affected by the vibrating object pass on these compressions and rarefactions to the adjacent air molecules, thus propagating the sound. The velocity of sound has to be determined by the speed with which the air molecules can pass on these compressions and rarefactions to adjacent air molecules. It stands to reason that the average velocity of air molecules must therefore be related to the velocity of sound and, in order to make up for randomness and collision effects, is probably higher than the velocity of sound. In these days of supersonic planes, many of us know that Mach 1, the speed of sound, is about 600 or 700 miles per hour, which comes out to be about 30,000 cm/sec. In earlier years we knew that if we counted the number of seconds between seeing lightning and hearing thunder and then multiplied by 1000, the result would be the number of feet between us and the lightning strike. This was because the speed of sound was 1000 feet/sec, which also works out to be around 30,000 cm/sec. The actual handbook value is 33,200 cm/sec (3.32×10^4 cm/sec).

Going back to the mean free path of a water vapor molecule in air (6.81×10^{-6} cm), this velocity of 3.32×10^4 cm/sec means that the time between collisions would be 2.05×10^{-10} sec, or that there are 3.87×10^9 collisions per second. Actually, the average velocity of an air molecule is 4.85×10^4 cm/sec, and the average velocity of a water molecule is 6.15×10^4 cm/sec so a water molecule experiences a collision every 1.11×10^{-10} sec, or 9.02×10^9 collisions/sec.

In the light of this knowledge we can reconsider our effort to evaporate water vapor from a pan of water in air in order to deposit a coating of frost onto a cold pane of glass suspended 10 cm above the pan. We gave that up because the evaporated water molecule would suffer so many collisions and would have to travel such a long, meandering path on the way that it seemed that it would never get from the pan to the pane. Now, in view of the speed with which gas molecules travel, it seems quite reasonable that water molecules can make their way from source (pan of water) to substrate (pane of glass) in reasonable times. We shall see that the process would be much more efficient in a vacuum; i.e., without the presence of air.

Evaporation in a Vacuum

We recall that we found that the diameter an air molecule is 3.74×10^{-8} cm. We also found that air in the gaseous state at normal temperature and pressure is mostly empty space with an average distance of 3.34×10^{-7} cm between air molecules. Finally, we found that the mean free path of a water molecule in air at normal temperature and pressure is 5.01×10^{-6} cm. We should remember that more accurate calculations using more sophisticated concepts give a mean free path of 6.81×10^{-6} cm. Here it will be better to use our crude result because we shall be making a number of such calculations and comparing one with another. The comparisons will be more meaningful with all results derived by similar methods.

Now we shall consider what would happen if we were to put our pan of water and cold pane of glass in an enclosed chamber and remove some of the air from the chamber. We shall have first removed the dissolved air from the water either by heating (and recooling) the water or by prior (and careful) evacuation. We shall start out by first removing half the air from the chamber. We can use the same ideas that we used previously to calculate that the distance between air molecules is 4.21×10^{-7} cm after half the air has been removed from the chamber. We can use the same ideas for calculating the risk that a water molecule might collide with an air molecule to find that the risk has been reduced to 1 chance in 48. We can make the guess that, on the average, a water molecule will travel through about 24 of our fictional layers (4.21×10^{-7} cm apart), a distance of 10.10×10^{-6} cm, before colliding with an air molecule. These crude methods of calculating the mean free path had previously given a mean free path of 5.01×10^{-6} cm under normal conditions. Removing half of the air has doubled the mean free path. Doubling the mean free path is not enough; we want to have the water molecule travel all the way from the pan to the pane unimpeded. To achieve this will require an increase of a factor of about 1×10^6 in mean free path. Removing half the air (therefore keeping half the air) has given us a factor of 2. We shall remove 90% of the air (keep 10%) and see what result this gives us. We find a distance of 7.20×10^{-7} cm between air molecules, a collision chance of 1 in 141, and a mean free path of $70.5 \times 7.20 \times 10^{-7}$ cm, or 5.08×10^{-5} cm. Reducing the amount of air to a tenth of the normal amount at normal temperature and pressure has increased the mean free path by a factor of 10. We have stumbled onto the fact that the length of the mean free path is inversely proportional to the amount of gas in the chamber. We actually saw this in one of our earlier expressions, MFP $= 1/2\pi n(r + r_t)^2$. The amount of gas in the chamber is n times the chamber volume. Since the mean free path is inversely proportional to n, it follows that it is inversely proportional to the amount of gas in the chamber. We are well aware that n is the factor of importance because this is what determines the number of gas molecules with which the moving molecule may collide. The higher the density of gas molecules, the greater the chance of collision is and therefore the shorter the mean free path.

We had said that we have to multiply the mean free path found at normal temperature and pressure of air by a factor of 1×10^6. To do this will require that we remove all but 1×10^{-6} the air from the chamber. The mean free path of a water molecule would then be 5.01×10^{-6} cm $\times 10^6$, or 5.01 cm. Since the distance from the pan of water to the cold pane of glass is 10 cm, the average water molecule would then make only two collisions in traveling from pan to pane. This, remember, is by our calculation. More sophisticated analysis finds that the mean free path would be 6.81 cm, and therefore the average water molecule would experience only one collision on the way. When we note the fact that our calculations have included mere brushes and elbow bumps between molecules as collisions, we can appreciate that a mean free path of 6.81 cm is a long mean free path. These light bumps do not noticeably deflect the water molecule, and it can travel distances of about ten times the mean free path before the deflections add up to significant turns or before suffering a real head-on collision. By evacuating the chamber, we have made it much easier to deposit coatings by evaporation.

We are beginning to see that the amount of gas in the chamber is a matter of much importance. More accurately, the number of gas molecules per cubic centimeter is the important quantity. Casting about for some item of familiarity with which we might associate this factor, we think of a gasoline service station. Many service stations have a facility for inflating tires. Most of us are aware that this facility includes a compressor to compress air and a chamber in which the air may be stored. We know that, as more air is compressed into the chamber, the chamber pressure is increased. We know also that, as we use the air from the chamber to inflate a tire, putting air from the chamber into the tire causes the tire pressure to increase. In both cases increasing the amount of air results in increased pressure. We should not be greatly astonished to hear that pressure is directly proportional to the amount of gas. If we double the amount of gas, we double the pressure. We have seen gauges which measure chamber pressure so we know it can be measured. From here on we are going to refer mostly to chamber pressure because it is a quantity that we can measure by use of a gauge. We shall remember, however, that the importance of pressure is that it is a direct measure of the amount of gas present, i.e., the number of gas molecules per cubic centimeter.

The concept of the mean free path of a molecule moving through gas will assume such importance in later sections that it would be wise to note here the dependence of the MFP on gas pressure. We know that the MFP is inversely proportional to gas density. We have seen here that gas density is directly proportional to gas pressure. It follows that the MFP is inversely proportional to gas pressure.

It is not apparent at this point that we should be interested in gases other than air, but we shall be. It is well, therefore, to pursue the question of the meaning of pressure for different gases. We can refer back to one of our earlier calculations, where we found that, the molecular weight of air being 29 and the weight of a nucleon being 1.66×10^{-24} cm, the weight of a molecule of air is 4.81×10^{-23} g. We concluded therefore that a gram of air must contain 2.08×10^{22} molecules of air. We found in a handbook that the density of air at normal temperature and pressure is 1.29×10^{-3} g/cm^3 from which we concluded that a gram of air occupies 775 cm^3. Since 2.08×10^{22} molecules of air occupy 775 cm^3, the numerical density of air is 2.68×10^{19} molecules of air per cubic centimeter at normal temperature and pressure.

We found for argon that, the atomic weight being 40, the weight of an argon atom is 6.64×10^{-23} g. From this we know that a gram of argon contains 1.51×10^{22} argon atoms. We can find a handbook value of 1.78×10^{-3} g/cm^3 for the density of argon at normal temperature and pressure from which we can conclude that a gram of argon occupies 562 cm^3. Since 1.51×10^{22} argon atoms occupy 562 cm^3, the numerical density of argon is 2.69×10^{19} atoms of argon per cubic centimeter at normal temperature and pressure. Comparing the numerical density of air (2.68×10^{19}) and of argon (2.69×10^{19}) we must conclude that these two numbers are too close to each other for it to be a matter of coincidence. If we made these calculations for other gases, we would find similar results. In fact, if we had been more accurate with our numbers and calculations, we would have found that all gases at standard temperature and pressure have a numerical density of 2.69×10^{19} molecules (or atoms) per cubic centimeter. This number is called Loschmidt's number after a person who happened to have stumbled onto it before we did. A more general statement, known as Avogadro's law, is that, under the same pressure and temperature, equal volumes of gases contain an equal number of molecules. We shall do well to keep this in mind because it is a profound result. Pressure has the same meaning for one gas as for another regardless of what gas pressure is at question. Let us keep this in mind for later sections in which we consider some very esoteric gases, namely, metal vapors.

CHAPTER II

VACUUM TECHNOLOGY

Our interest in pressure, especially reduced pressure, is established. We should like to have a pressure of zero, but this is impossible. It is even very difficult to get pressures as low as those we can get by with; we shall learn to appreciate the ingenuity of those who have blazed the trail in vacuum technology.

It is not likely that any of us have any doubt as to how we are going to get a vacuum in our vacuum chamber. We know that we are going to use a vacuum pump, but now that we are ready to consider this matter we need to know exactly what a vacuum pump is. As a starting point, let us refer again to the gasoline service station with an air compressor and tank. The air compressor takes in air from the atmosphere, compresses it to a higher pressure, and exhausts it into the compressed air tank. What would happen if the air compressor were connected backwards? If the intake of the compressor were connected to the tank and the output left open to the atmosphere, the compressor would take in air from the tank (thus reducing the tank pressure), compress the air, and exhaust the air into the atmosphere. By the blunder of connecting the air compressor backwards, we have converted it to a vacuum pump, and we have converted the tank to a vacuum chamber.

Mechanical Vacuum Pump

An air compressor connected backwards would not be a very efficient vacuum pump. Air compressors usually are piston-type compressors, which are difficult to use as vacuum pumps. Vacuum pumps (more specifically, mechanical vacuum pumps) are usually of the oil-sealed rotary design illustrated in Figure 1. The axis of rotation of the rotor is above the axis of the cylindrical bore of the stator by an amount equal to the difference in radii. Thus the top of the rotor seats against the top of the cylindrical bore of the stator closely enough that the lubricating oil will form a vacuum seal at the junction line. The bore, the rotor, their junction line, and the two sliding vanes form three volumes. Volume 1, connected to the inlet, is expanding. Gases in a vacuum chamber connected to the inlet can expand into volume 1 and are thus removed from the chamber. Volume 2 contains gas that has been removed from the vacuum chamber and is being compressed. Volume 3, connected to the exhaust through a flapper valve, contains gas being compressed further. When the pressure in volume 3 reaches atmospheric pressure, the gas in volume 3 can escape through the flapper valve and out the exhaust to the atmosphere. When the

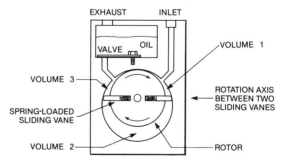

FIGURE 1 Schematic representation of a common type of mechanical pump.

pressure in volume 3 is below atmospheric pressure, the flapper valve prevents air from leaking backwards from the atmosphere into volume 3.

When the vane separating volume 3 from volume 2 has passed the line connecting volume 3 to the exhaust, there is a small amount of gas entrapped. Most of this gas, as it is further compressed, is able to leak through the vacuum seal between the spring-loaded vane and the wall of the cylindrical bore of the stator and back into the new volume 3 (formerly volume 2). Some of this gas leaks into volume 1 through the vacuum seal between the rotor and the bore and thus puts limits on both the pumping speed and the ultimate vacuum that can be achieved. These limits can be alleviated by using two stages in series, and most vacuum pumps are manufactured as two-stage pumps. The second stage, maintaining the exhaust of the first stage at a low pressure, leaves very little gas to leak back into the inlet volume of the first stage. One might think that using more stages in series would lead to a higher vacuum (lower pressure), but other factors become important so that two stages are the practical limit.

Pumping Speed of a Mechanical Pump

We are interested in two aspects of the performance of this mechanical vacuum pump: how quickly it can evacuate a chamber, and how low a pressure it can attain. We see that it removes air from the chamber in gulps having volumes equal essentially to volume 2 in Figure 1. It takes two gulps per revolution of the rotor. We can find that electric motor speeds are typically about 1725 RPM. We can find that typical pulley ratios are in the range 4 or 5 to 1. We are therefore not too surprised to find that typical pump rotor speeds are near 400 RPM so that air is being removed from the vacuum chamber at a rate of 800 gulps per minute. If volume 2 is a half liter, then the pumping speed is 400 liters/min. Evidently volume 2 on most of our pumps is a little bigger than this because they are running at a little less than 400 RPM and yet they have pumping speeds of about 500 liters/min.

We now want to see if we can develop an understanding of

pumping speed. We shall consider an exaggerated case in which our vacuum chamber has a volume of 1 liter (1000 cm³), our pump has a volume 2 of 1 liter, and we can check the pressure after each half turn of the pump rotor. We recall that we found that 2.08×10^{22} molecules of air occupy a volume of 775 cm³. From this we find that 1 cm³ of air contains 2.68×10^{19} molecules of air; thus 1 liter of air contains 2.68×10^{22} molecules of air. In each half turn of the pump rotor, volume 2 has been connected to our vacuum chamber. The speed of gas molecules is so great that the air in the chamber has filled both the chamber and volume 2 of the pump completely during the time of a half turn of the pump rotor. Thus half the air originally in the chamber remains in the chamber and half enters volume 2. In this way, at each half turn of the pump, half the air is removed from the chamber and half remains in the chamber. At the start the chamber, at atmospheric pressure, contains 2.68×10^{22} molecules of air. After a half turn of the pump the chamber pressure is reduced to a half atmosphere, and the chamber contains 1.34×10^{22} molecules of air. In the next half turn of the pump a new volume 2 has been exposed to the chamber, and the air remaining in the chamber again fills both the chamber and volume 2. Again half of the air in the chamber at the start of the half turn of the pump remains in the chamber and half enters volume 2 to be removed. After the second half turn the chamber pressure is down to a fourth of an atmosphere and the chamber contains 6.7×10^{21} molecules of air. We can deduce that at each half turn the chamber pressure and the number of air molecules in the chamber will be reduced by a factor of 2. After a number of half turns, which we designate by the letter n, the chamber pressure and the number of air molecules in the chamber will be reduced by a factor of 2^n.

We recall that we found that the chamber pressure would have to be reduced by a factor of 10^6 in order to have an acceptably long mean free path. We are interested in the value of n for which 2^n equals 10^6. Using a scientific calculator we can quickly find by trial and error that $n = 20$ is about it, but, for devious reasons, we shall do it here the hard way. We make the equation $2^n = 10^6$ and take logarithms to get $\log 2^n = \log 10^6$. This gives us $n \log 2 = 6 \log 10 = 6$, from which we get $n = 6/\log 2 = 6/0.301 = 19.93$ in agreement with our trial-and-error result.

Using the same method to investigate the possibility of getting down to zero pressure we find that this would require an infinite number of half turns. This is not the source of the statement at the beginning of this chapter that getting zero pressure is impossible. We can find the "exact" number of half turns to get the pressure down low enough that we have only about three molecules left in the chamber. This is achieved by the number of half turns for which $2^n = 10^{22}$ or $n = 73$ half turns. At 800 half turns/min we get there in about 5 sec. Clearly there are other reasons that zero pressure is impossible, but before we pursue the question of ultimate (low) pressure or vacuum we shall try to deduce more with respect to pumping speed.

We are interested in estimating how much time is required to evacuate a vacuum chamber. We can reasonably assume that the time t required should be proportional to the chamber volume V; i.e., $t = k_1 V$, where k_1 is a constant of proportionality. A chamber twice as big should take twice as long; one that is half as big, half as long. We can reasonably assume that the time required should be inversely proportional to the pumping speed S of the pump; i.e., $t = k_2/S$, where k_2 is a constant of proportionality. A pump twice as big should take half the time; one that is half as big, twice as long. The difficulty is in seeing how the pressure is involved. We know that as the chamber pressure goes down the pump is able to take out less gas. In our example above the pump took out 1.34×10^{22} molecules of air in the first half turn but only 6.7×10^{21} in the second half turn, and in the 72nd half turn it removed only three molecules. These reductions are due to the chamber pressure going down as the pumping continues. This is the factor we want to include. Our reason for using the hard way to calculate the number of half turns required to reduce the chamber pressure was to introduce the pressure factor. The 10^6 was exactly the ratio of the initial pressure p_0 to the final pressure p; i.e., p_0/p. Had we needed some ratio other than 10^6, we would have calculated that ratio and used it instead of 10^6. We would have made the equation $2^n = p_0/p$ and taken logarithms to get $\log 2^n = \log(p_0/p)$, or $n \log 2 = \log(p_0/p)$. The number of half turns is proportional to time t; i.e., $n = k_3 t$, where k_3 is a constant of proportionality. Since $n = (\log 2)^{-1} \log(p_0/p)$, we can see that the pumping time is proportional to $\log(p_0/p)$; i.e., $t = n/k_3$, or $t = (k_3 \log 2)^{-1} \log(p_0/p)$. We now have all the factors involved in calculating pumping time, $t = (kV/S)$ $\log(p_0/p)$, where k is a constant of proportionality incorporating k_1, k_2, k_3, and log 2. More precise derivation gives the result:

$$t = \frac{2.3V}{S} \log \frac{p_0}{p},$$

where t is the time required to pump a chamber of volume V from an initial pressure p_0 to a final pressure p by use of a pump having a pumping speed S. A reasonable chamber volume might be about 200 liters, a reasonable pump speed might be about 500 liters/min, and a reasonable pressure ratio might be about 10^6, so a pumping time of about 6 min is reasonable. Each minute of pumping time has gained an order of magnitude in the vacuum, and we might thus deduce that we can achieve whatever vacuum we might choose by pumping an appropriate number of minutes. In practice we find that pumping does not go that well. The first four orders of magnitude below atmospheric pressure follow this equation quite closely. The fifth slows down, and the sixth drags practically to a stop. Leakage of air together with dead volumes within the body of the pump mechanism limits the ultimate pressure ratio obtainable with a mechanical vacuum pump. Manufacturer's specifications actually indicate an ultimate pressure one order of magnitude lower, and this is attainable but not in a practical sense.

Pressure Measurement

Now that we have a vacuum pump and some indication of its performance, we want to investigate the problem of measuring the pressures attained. Dial-type pressure gauges are so common that most of us have seen them somewhere at some time. These gauges are based on the mechanical displacement of some sort of metallic diaphragm (usually, although not always, coiled) caused by forces resulting from pressure differentials. The pressure on one side of the diaphragm is usually atmospheric pressure. In these days of constant weather reports we know that atmospheric (barometric, in the language of the weather reports) pressure is continually changing so the reliability of readings from these gauges immediately becomes suspect.

Atmospheric pressure is a good subject with which to introduce our units of pressure measurement. Many of us may not remember, but it is unlikely that any of us went through school without having heard how Torricelli made his first barometer. He used a glass tube longer than 760 mm (millimeters) closed at one end and open at the other. If the diameter of this tube was large enough, it looked like a big test tube. The chances are it had a small diameter, but, as long as he had enough mercury to fill the tube, the diameter does not really matter too much. After completely filling the tube with mercury, Torricelli inverted it and inserted the open end in a basin of mercury. After enough tries he was able to do this without letting air into the tube, but even then there was always an empty space above the mercury. The column of mercury was always approximately 760 mm high; the rest of the tube above the mercury level was empty, regardless of the length or diameter of the tube. It was found that the height of the column of mercury averaged about 760 mm, went up to 770 or 780 mm in fair weather and went down to 740 or 750 mm in cloudy weather. Torricelli had a simple instrument that measured atmospheric pressure accurately. It was found that the same result could be achieved without the basin of mercury if the glass tube were bent into a U-tube shape with vertical legs longer than 760 mm.

Since the height of the mercury column above the height of the mercury pool is the measure of atmospheric pressure, it became the custom to speak of pressure in terms of millimeters of mercury (mm Hg). Our blood pressure is measured by a column of mercury, and the numbers quoted by the doctor are actually millimeters of mercury. When people began to measure lower pressures, the U-tube adaptation was a convenient pressure gauge, and pressure measurements in units of millimeters of mercury became the rule at low pressures. We continue this practice today in vacuum pressure measurements, but we have now changed the name from millimeters of mercury to torr. It is not too difficult to guess the name of the man who has thus been honored.

It will come as no surprise that normal atmospheric pressure is considered to be 760 mm Hg or, in our up-to-date terminology, 760

Torr. We recall that we had come to the conclusion that, in order to have a long enough mean free path, we wanted our chamber pressure to be decreased by a factor of 10^6. We want our chamber pressure to be 760 Torr $\times$ 10^{-6} or 0.76×10^{-3} Torr. We encounter vacuums in the range of 10^{-3} Torr so regularly with mechanical pumps that we habitually use the prefix "m" (milli) to substitute for 10^{-3}. We thus speak of a pressure such as 0.76×10^{-3} Torr as 0.76 mTorr. Those of us who are older or who have much associated with older scientists usually speak of a micron instead of a millitorr. This term arises from the term millimeter of mercury. A millimeter is a thousandth of a meter. A thousandth of a millimeter is a millionth of a meter. The prefix for millionth is micro. The origin of the "n" to make it micron is a mystery. We shall endeavor to say torr and millitorr.

We have concluded that we want our chamber pressure to be a millionth of atmospheric pressure or 0.76 mTorr. For all practical considerations, 1 mTorr is equally good, and, since 1 is easier to say and to write, we shall now consider this to be the pressure at which our mean free path will be long enough that particle collisions will not be a problem.

Looking at a scale divided into units of centimeters and millimeters, one can readily see that the Torricelli barometer, or the U-tube adaptation thereof, is not highly useful in measuring pressures below 1 mm Hg; i.e., 1 Torr. At a millitorr, such an instrument has become totally incapable of yielding meaningful results. Since dial pressure gauges extract their movement from forces resulting from pressure differentials, it is not too surprising that these too cease to function significantly below 1 Torr.

Thermocouple Vacuum Gauge

A pressure gauge that can be used in the millitorr range (1–1000 mTorr) is the thermocouple vacuum gauge. We can describe this gauge and its operation, but it cannot be easily related to common knowledge or common experience. This gauge is referred to as a vacuum gauge rather than as a pressure gauge because we have come to refer to pressures below 1000 mTorr (1 Torr) as a vacuum.

A schematic sketch of a thermocouple vacuum gauge is shown in Figure 2. The pressure sensing element consists of a heater wire to the center of which a thermocouple junction has been spot welded. The source of power for the heater wire is a 6-V (volt) transformer. The heater-wire current is adjusted, with the gauge tube at a vacuum appreciably higher (i.e., pressure appreciably lower) than 1 mTorr, to the value for which the readout (a millivoltmeter) reads full scale (14 mV). Under these vacuum conditions, heat is being removed from the wire only by radiation and by conduction along the wires to the tube base. The total amount of power involved is less than a watt so the tube envelope and base easily maintain ambient temperature and the reference junction of the thermocouple is thus essentially at room temperature. Gas molecules that impinge against the heater wire acquire higher velocities consistent with the wire temperature (i.e., are heated), and their next collision, being against some part of the gauge tube envelope, transfers the energy acquired from the wire to the tube wall. Under these vacuum conditions, this cooling effect is not detectable by the readout gauge. As far as the readout gauge is concerned, the pressure is zero, and the full-scale reading is calibrated as zero. As gas is admitted to the vacuum system, this cooling effect remains undetectable until the pressure reaches 1 mTorr. At 1 mTorr this cooling effect reduces the temperature of the wire enough that a

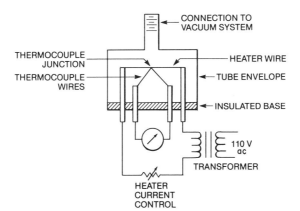

FIGURE 2   Thermocouple vacuum gauge.

decrease in the gauge reading is detected. As more gas is admitted, this cooling effect increases, and the gauge reading decreases. As further gas is admitted, this cooling effect reaches a maximum at about 1000 mTorr (1 Torr), and subsequent addition of more gas does not change the gauge reading enough to be calibrated. Thus this gauge is effective in the range 1 to 1000 mTorr. At the high pressure end, this gauge starts in where U-tubes and mechanical dial gauges leave off. At the low pressure end, this gauge stops at the same limit as our mechanical pump (1 mTorr). We recall that 1 mTorr is also the pressure (or vacuum) at which collisions with gas molecules will no longer interfere with evaporated atoms.

Impacts of Gas Molecules against Solid Surfaces in a Vacuum Chamber

There have been vague indications that a vacuum of 1 mTorr may not be sufficient for our evaporative coating process. We now want to pursue this question. We recall that the mean free path in air at normal temperature and pressure is about 5×10^{-6} cm. This means that the average molecule can travel only 5×10^{-6} cm before it makes some kind of a collision with an air molecule. By the time the average molecule has traveled about 50×10^{-6} cm it has experienced a major change in its direction of travel. The general motion of the average molecule in air at normal temperature and pressure is thus governed by other air molecules. We recall that we found that the length of the mean free path is inversely proportional to pressure. If we reduce the pressure by a factor of 10, we increase the mean free path by a factor of 10, to 50×10^{-6} cm. Always estimating ten mean free paths as the length of path an average molecule travels to have been deflected significantly, the average molecule can now travel 500×10^{-6} cm. Let us coin the term mean deflection path for this distance; i.e., the mean deflection path is equal to ten times the mean free path. A mean deflection path of 500×10^{-6} cm is still so short that the general motion of the average molecule is still governed by other air molecules.

Let us consider the situation at a pressure of 1 mTorr. Here the mean free path is about 4 cm, and the mean deflection path is about 40 cm. We can see here a useful relationship. If the gas pressure is given in millitorr, then the mean free path in centimeters is given by MFP = $4/p$. This expression is reasonably valid for any atomic or molecular particle with which we shall deal moving through any gas. In the present case with $p = 1$ we have MFP = 4 cm, and a mean deflection path of 40 cm. The average molecule can travel from one wall of the vacuum chamber to another without experiencing a significant collision with another gaseous molecule. The average molecule bounces from wall to wall in the chamber almost as if it were the only molecule in the whole chamber. The general motion of the average molecule at a pressure of 1 mTorr has become totally independent of the presence of other air molecules.

Referring back to the description of the operation of the thermocouple vacuum gauge, this effect is closely related to the fact that there is a high pressure limit for this gauge. At 1 mTorr a gas molecule that has acquired energy from the heated wire will make its next impact with the gauge tube envelope. At somewhat higher pressures gas molecules will make a collision or several collisions with other gas molecules and transmit some of this energy to them before reaching the tube wall to give up the rest of the excess energy. At 1000 mTorr and higher pressures a gas molecule that has acquired energy from the heated wire will have passed on all of its excess energy to other gas molecules before reaching the tube wall. Under these conditions heat transfer through a gas is independent of pressure. Heat transfers through gas at the same rate at 1000 mTorr as at 1 atmosphere or as at 100 atmospheres.

If we remember that the chamber walls are made up of atoms and molecules, we can appreciate that the wall surfaces look different to gas atoms than they look to us. We may see the walls as smooth and highly polished. As seen by gas atoms and molecules, wall surfaces are composed of mountains and valleys, crags and ravines, boulders and craters, all made up of clusters of huge rocks. The direction in which a molecule will rebound after impinging against the wall depends on which side of the mountain or valley it strikes and even depends on which side of the rock it strikes. Gas molecules in a chamber at a pressure of 1 mTorr (or less) therefore bounce around from wall to wall in random directions and can be pumped out only as they, by chance, happen to bounce into the port of the chamber to which the pump is attached.

We are going to see what we can learn by making some very crude calculations of the motions of gaseous molecules in a vacuum chamber at a pressure of 1 mTorr or less. Vacuum chambers are seldom built as cubes of dimension l or as spheres of diameter l, but neither are they built with any one dimension greatly different from any other. We are going to say that the average linear dimension of our chamber is l. We are going to say that the inside surface area is $4l^2$. If it were a hollow cube, its inside surface area would be $6l^2$, and the inside surface area of a hollow sphere would be πl^2 so $4l^2$ cannot be too far off. We are going to say the volume of our chamber is l^3, which would be exactly correct for a hollow cube and is not too far off for any other shape. We have calculated the number of air molecules per cubic centimeter at various pressures, but for the general case we shall denote this quantity as n. The total number of gas molecules in the chamber is thus nl^3. We made a good guess at the average velocity of an air molecule, and we looked up the exact value in a handbook, but we shall now simply denote this as v. If we denote the time required for the average gas molecule to traverse the average path across our vacuum chamber by τ (the Greek letter tau), then we have $\tau = l/v$. We know that v is very large so that τ is very small; therefore each gas molecule is going to impinge against the chamber walls many times per second. If the number of times per second each gas molecule impinges against some part of the chamber wall is denoted by N_0, then we have $N_0 = 1/\tau = v/l$. Since there are a total of nl^3 gas molecules in the chamber, then the chamber walls suffer a total of $nl^3 N_0$ or $nl^2 v$ collisions per second by gas molecules. Since the total chamber wall area is $4l^2$, then each square centimeter of chamber wall will experience $nl^2 v/4l^2$, or $\frac{1}{4} nv$ collisions per second. This is exactly the result obtained using the most exact and elegant derivation. This is a profoundly important result. We use it in later sections to calculate pumping speeds and also to analyze the phenomenon of evaporation. We use it here to come to an understanding of the problem of surface contamination.

We remember that we found that we have $n = 2.68 \times 10^{19}$ air molecules/cm^3 at normal temperature and pressure (NTP) and that the amount of air in the chamber or the density of air or the numerical density of air is proportional to pressure. Since our chamber is at 1 mTorr, we have

$$n = \frac{2.68 \times 10^{19}}{760,000} = 3.53 \times 10^{13} \quad \text{air molecules/cm}^3.$$

Remembering that $v = 4.85 \times 10^4$ cm/sec, we find from $\frac{1}{4}nv$ that each square centimeter of chamber wall experiences 4.28×10^{17} collisions/sec from air molecules. This is true not only for the chamber wall but also for any other surface inside the chamber. If we now remember that all atoms are about the same size, 2.5×10^{-8} cm, and cover about the same area, $(2.5 \times 10^{-8}$ cm$)^2$ or 6.25×10^{-16} cm^2, in a solid surface, then we find that each atom of a surface in a chamber at 1 mTorr experiences $4.28 \times 10^{17} \times 6.25 \times 10^{-16} = 267$ collisions/sec.

It is not easy to see the importance of this. We know from fairly common experience that many materials, especially metals, oxidize when exposed to heat and air. We know that air contains about 20% oxygen and 80% nitrogen. We know that we consume oxygen as we breathe and that nitrogen gas is inert with respect to biological interactions. We are so accustomed to thinking of nitrogen as being an inert gas biologically that we think of it as being inert in all respects. This is not the case. We can look in the tables of physical properties of inorganic compounds in the "Chemical Rubber Company Handbook of Physics and Chemistry" and find nitrides of almost all metals. Materials that oxidize when exposed to heat and air will also nitridize. In fact, heat is not a necessary ingredient. The contribution of heat is to speed the penetration of the reactions to deeper layers. This penetration may be by diffusion of gases through reacted layers to reach unreacted layers, by exchange between reacted layers and unreacted layers, or by physical breakdown of reacted layers to allow gases to reach unreacted layers. Without heat, these reactions (oxidation and nitridation) will occur instantly at the surface, but will take much longer to penetrate to deeper layers. The important fact is that these reactions will occur instantly at the surface without need for heat or high temperature.

In view of this, the 267 collisions/sec experienced by a surface

atom in the presence of gas at 1-mTorr pressure become significant. As we deposit a coating, these impinging gas atoms will react
with the deposited atoms, yielding a contaminated coating. We
consider this in more detail in later sections, where we shall find
that this contamination problem can be reduced to negligible pro-
portions only if the vacuum is improved by many orders of magnitude. Our vacuum of 1 mTorr, the practical limit for this type of
mechanical vacuum pump, turns out to be woefully inadequate.
We need a much more efficient vacuum pump.

Diffusion Pump

We are now in the same situation we were in when we had to
introduce the thermocouple vacuum gauge. We have to introduce
the diffusion pump without any preparatory discussion. A schematic sketch of a diffusion pump is shown in Figure 3. The operation of a diffusion pump is very crudely analogous to the effect of a
jet of air on a cloud of tobacco smoke. Most of us have at some time
in our lives been in the vicinity of a puff·or small cloud of tobacco

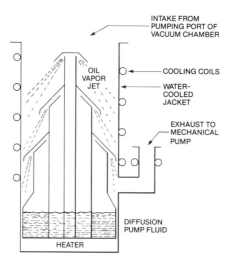

FIGURE 3 Schematic representation of a diffusion pump.

smoke, which, when blown at, can be caused to move. The jet of air blown at the cloud entrains some of the smoke as it passes through the cloud, and the smoke can thus be caused to move away. The entrainment is the analogous factor. No other similarity exists. In the diffusion pump, gas molecules are pumped by being entrained in oil vapor jets.

We remember that, in the pressure range from 1 mTorr down, gas molecules bounce from wall to wall of the vacuum chamber without ever colliding with other gas molecules. In this pressure range, gas molecules can be pumped out of the chamber only when their bouncing happens to be such that they, by chance, hit the pumping port. It occurred to W. Gaede, something like 70 years ago, that a jet of vapor laying in wait at the pumping port could entrain these gas molecules as they came through the pumping port, add a directed motion to their random motion, and prevent them from reentering the vacuum chamber. With the jet directed away from the chamber and toward the mechanical pump, this directed motion enhances the pumping efficiency of the system. Some modifications over the years resulted in the pump shown schematically in the sketch in Figure 3. Gaede had used mercury as the diffusion pump fluid, but within a few years it was found that certain oils were more effective under many conditions. Over the years a number of highly suitable oils have been developed as diffusion pump fluids, and mercury is now used only in highly specialized applications. With the system pumped to the low milli-torr range, the diffusion pump oil in the bottom of the pump is heated to a temperature (200 to 250°C) at which it has a vapor pressure of about 500 mTorr. The interior of the conical sections of the pumping stack is thus filled with oil vapor at this pressure. The oil vapor expands out from the interior of the stack through the conical ports and is directed downward toward the pump exhaust by the conical deflectors.

Any gas molecule that (by random motion) enters the first jet of oil vapor is given a directed motion downward that enhances the probability that it will enter the second jet. The second jet further confines the entrained gas molecules and adds more directed motion toward the exhaust. The third jet confines the gas molecules so that they can be pumped out by the mechanical pump connected to the exhaust. The oil vapor impinges on the water-cooled wall of the diffusion pump jacket where it condenses and drains down to the bottom to be recycled. Gas molecules that impinge on the wall rebound and, being entrained by the oil vapor jet, continue in general motion toward the exhaust.

We may wonder if the fact that the gas molecules leave through the pumping port only by chance does not result in very low pumping speeds. We remember, however, that the number of air molecules impinging on the chamber walls (or passing through the pumping port) per square centimeter every second is very large. At a pressure of 1 mTorr this number is $(4.28 \times 10^{17}/cm^2)/sec$. This number, we recall, is proportional to pressure, so if the pressure is reduced to 0.5 mTorr, then the number of air molecules impinging per square centimeter each second drops to 2.14×10^{17}. In general, if the pressure p is given in millitorr, then (4.28×10^{17}) p gas molecules per square centimeter leave the pumping port by random motion each second. If A is the area of the pumping port, then the total number of gas molecules leaving the pumping port each second is (4.28×10^{17}) pA. We recall that the number density of gas molecules in a chamber at a pressure of 1 mTorr is 3.53×10^{13} gas molecules/cm^3. We remember that this number is also proportional to pressure. If the pressure is doubled, the number is doubled; if the pressure is halved, the number is halved. If the pressure is given in millitorr, then a chamber at pressure p contains (3.53×10^{13}) p gas molecules/cm^3. We know that, at a pressure p, (4.28×10^{17}) pA gas molecules will leave through a pumping port of area A each second and that these gas molecules would occupy a volume of $1/(3.53 \times 10^{13})$ p cm^3. The volume rate at which gas leaves the chamber is thus (4.28×10^{17}) $pA/3.53 \times 10^{13})$ p, or $12,000A$ cm^3/sec. A typical vacuum chamber may have a pumping port of 15-cm diameter, or an area of 180 cm^2, which results in a pumping speed of 2,160,000 cm^3/sec. This is more commonly given in liters per second; i.e., 2160 liters/sec. Recalling that a typical mechanical pump will have a pumping speed of about 500 liters/min, or about 8 liters/sec, we see that Gaede's idea for enhancing the speed at which a vacuum chamber may be pumped was quite ingenious.

Strictly speaking, we have calculated not the pumping speed of the diffusion pump but the pumping speed of the pumping port of the vacuum chamber or the pumping speed of the entrance port of the diffusion pump. These ports are normally of the same size. This pumping speed can be thought of as the theoretical maximum pumping speed. If the oil vapor jets are not 100% efficient in entraining gas molecules that come through the port, then the pumping speed will be less than this limit. Even if a diffusion pump were 100% efficient, it is not possible to achieve the maximum pumping speed at the vacuum chamber. The pumping port of the vacuum chamber is seldom connected directly to the entrance port of the diffusion pump. We shall see later that there are valves, cold traps, and interconnecting tubing between the pumping port of the chamber and the entrance port of the pump. These components impede the motion of gas molecules from the chamber to the pump and thus reduce the total effective pumping speed. Nevertheless, the diffusion pump provides orders-of-magnitude improvement in efficiency and in the ultimate achievable vacuum.

A diffusion pump alone cannot achieve the vacuum levels we require. Mechanical pumps and diffusion pumps are complementary components of a vacuum system, and both are vital in achieving the desired vacuum levels. Diffusion pumps cannot operate at pressures above 1000 mTorr so it is necessary to use a mechanical pump in conjunction with the diffusion pump. The mechanical pump can remove the bulk of the gas from the chamber and bring the system pressure down to the range in which the diffusion pump can operate. A diffusion pump can operate for brief periods of time in the range 100–1000 mTorr under the proper circumstances, but it is a struggle at best. A diffusion pump can operate surprisingly well in the range 10–100 mTorr under the right circumstances, but this is seldom the most effective pump in this range. A diffusion pump is often highly effective in the range 1–10 mTorr, but is most useful and effective in the range from 1 mTorr down. A mechanical pump is always required: first to pump the system down to pressures at which the diffusion pump is effective and subsequently to remove the gases that are pumped from the vacuum chamber by the diffusion pump and exhausted into the foreline.

Ionization Gauge

We might now consider the question of how we are going to measure these higher vacuums (lower pressures). We have become inured to the introduction of new concepts so it comes as no great surprise to see another new instrument depicted in Figure 4. This form of ionization gauge, known as the Bayard–Alpert gauge, is the one now most commonly used for measuring pressures in the high vacuum range, i.e., below 1 mTorr. This gauge is based on ionization of residual gases in the vacuum chamber. Actually, it is the gases in the ion gauge tube that are subject to ionization, but it is assumed that the rapid motion of the gas molecules is such that the gases in the tube are an exact sampling of the gases in the chamber.

Ionization is caused by electrons that are emitted by a filament and accelerated toward the grid. Electrons traveling through gas will collide with gas atoms just as gas atoms traveling through gas will collide with other gas atoms. In such a collision the electron will give up energy to an orbital electron of the atom. If the energy

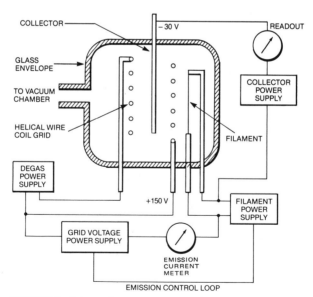

FIGURE 4 Schematic representation of an ionization gauge.

exchange is less than the ionization energy, the orbital electron is excited to a higher energy level for about 10^{-8} sec and immediately returns to the ground state with the emission of a photon of light. If the energy exchange is greater than the ionization energy, the orbital electron separates from the atom leaving a free electron and an ion. The ion may be attracted to either the filament or the collector depending on where the collision occurred in the tube.

Electrons are emitted by the filament when it is heated to a sufficiently high temperature. The filament power supply provides an ac voltage of near 6 V to heat the filament to emission temperature. The grid voltage power supply furnishes a constant $+150$ V to the grid to accelerate the electrons from the filament toward the grid. This electron emission current, usually 10 mA, is measured by the emission current meter and is held constant by an emission control loop between the grid power supply and the filament power supply. The helical wire coil, that comprises the grid is an open structure of ½-mm wire with ½-cm spacing and a diameter of 2 cm. As a consequence, the electric field is very nonuniform, and the electric field lines are curved. The electrons are unable to follow the electric field lines all the way to the grid wire, and many electrons penetrate the interior of the winding. The collector inside the grid coil causes a generally similar electric field back toward the grid coil forcing the electrons back out. Again the electrons are unable to follow the electric field lines all the way to the grid wire, and many electrons penetrate the region outside the grid wire. Thus many electrons cycle in and out giving them a long path of travel among the gas molecules before ultimately reaching the grid wire. This long path resulting from the electrons constantly missing the wire is very important because the mean free path of an electron in the gas is about six times the mean free path of a gas molecule. It is necessary to have a very large number of electrons traveling these long paths in order to have enough collisions with gas molecules that the resulting ionization is measurable. When a gas molecule within the interior of the helical wire coil grid is struck by an electron such that it is ionized, the resulting ion is attracted toward the central wire (appropriately referred to as the collector) because of the negative potential applied by the collector power supply. We have been down similar paths often enough that we recognize that the resulting ion current is proportional to pressure and the readout can thus be calibrated to read in pressure units.

Our mechanically actuated pressure gauges had been calibrated in units of 1 to 760 Torr. We had calibrated our thermocouple vacuum gauges in units of 1 to 1000 mTorr. In calibrating our ionization gauges we revert back to torr. We use our ion gauges in the range 10^{-3} to 10^{-7} Torr. The gauge is reliable down to 10^{-10} Torr, but we find that our system limit is at 10^{-7} Torr. There is valid reason to believe that most of our work is actually done in the range below 10^{-7} Torr, but the ion gauge reading will seldom be below 10^{-7} Torr. This anomaly is discussed in a later section.

Pumping Limits, Baffles, and Cold Traps

We now want to consider some of the factors that limit our system vacuum. If we look at the vapor pressure curves of Figure 1 of Chapter I, and try to estimate the temperature at which the vapor pressure of some material becomes zero, we conclude that the vapor pressure may become incredibly small but will never become zero for any material. We can be quite certain that the same is true for diffusion pump fluids, and, in fact, at room temperature we can find that vapor pressures for commonly used diffusion pump fluids are in the range 10^{-8}–10^{-10} Torr. Thus diffusion pump fluids directly set a limit on the ultimate vacuum. In addition, we are aware that oils, such as cooking oils and motor oils, are subject to a certain amount of thermal decomposition. We can be sure that diffusion pump fluids too are subject to thermal decomposition and that some of the breakdown products, being gaseous, can enter the vacuum chamber. Thus diffusion pump fluids indirectly set a limit on the ultimate vacuum. The diffusion pump oil vapor and breakdown products are included among gases that are referred to as backstreaming gases. The mechanical pump oil also has vapor and breakdown products that can work their way backward through the diffusion pump to the vacuum chamber. Air too can contribute to the backstreaming by working its way backwards through the mechanical pump and thence backwards through the diffusion pump into the vacuum chamber.

The backstreaming oil vapor and breakdown products are reduced by interposing water-cooled baffles between the diffusion pump and the vacuum chamber. The construction of these baffles must be such that no gas molecule can travel directly from the diffusion pump to the pumping port. A gas molecule from the diffusion pump should be prevented from reaching the pumping port without at least one collision (preferably two) with some section of the baffle wall. Such a baffle design is spoken of as being optically dense because light too can get through only by multiple reflections.

It is quite common practice to insert a liquid-nitrogen-cooled baffle between the water-cooled baffle and the vacuum chamber.

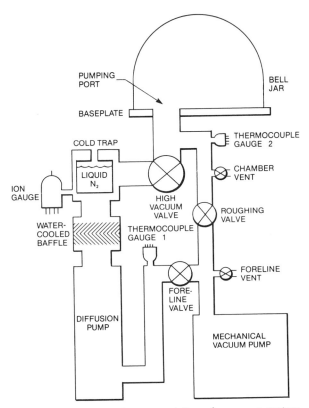

FIGURE 5 Schematic representation of a vacuum system.

The liquid-nitrogen-cooled baffle, more commonly referred to as a cold trap, is usually not an optically dense baffle and is frequently just a tank that is reentrant into an enlarged section of the tubing between the water-cooled baffle and the vacuum chamber. A cold trap of this type is shown in the depiction of a vacuum system in Figure 5. The backstreaming oil and decomposition products that get past the water-cooled baffle are reduced many orders of magnitude in pressure by the liquid-nitrogen-cooled cold trap. In addition, many gases that are not affected at the temperature of cold water are reduced many orders of magnitude in pressure by the liquid-nitrogen-cooled cold trap. Normally, one speaks of the cold trap, and it is implicitly assumed that it is cooled by liquid nitrogen. Similarly, if one speaks of a baffle, the implicit assumption is that it is water cooled. The gases that are strongly reduced by the cold trap are commonly referred to as condensable even though everyone is well aware that this is a misnomer.

A Vacuum System

A typical vacuum system, such as that depicted in Figure 5, is composed of three major groupings connected to each other or isolated from each other by valves. The vacuum chamber, which is discussed in the next section, has an attached thermocouple gauge. The diffusion pump with its baffle, cold trap, ion gauge, and thermocouple gauge is connected to or isolated from the vacuum chamber by the high vacuum valve. The mechanical pump is isolated from or connected to the vacuum chamber by the roughing valve. The mechanical pump and the diffusion pump are connected to or isolated from each other by the foreline valve. Vent valves provide for letting the vacuum chamber and the mechanical pump up to atmospheric pressure.

When the vacuum chamber is to be let up to atmospheric pressure, it is isolated by closing the high vacuum valve. It is implicitly assumed that the roughing valve is already closed because it is normally closed. Opening the chamber vent valve admits air to the vacuum chamber so that the bell jar can be raised permitting access to the working area of the chamber. When it is desired to evacuate the chamber, the foreline valve is closed to protect the diffusion pump. The roughing valve is then opened to rough out the chamber, a process that normally requires a pumping time of about 5 min. When thermocouple gauge 2 indicates that the chamber pressure is low enough (below 50 mTorr) to begin opening the high vacuum valve, the roughing valve is closed and the foreline valve is opened. Attention is then turned to thermocouple gauge 1, which monitors the foreline pressure. The term foreline has come to be used in referring to the pumping line connecting the diffusion pump exhaust to the mechnical pump (often spoken of as the forepump). The high vacuum valve should be opened carefully so as to limit the forepressure to values below 100 mTorr. As the high vacuum valve begins to open, the forepressure will be seen to rise appreciably. We recall that the pumping speed of the diffusion pump is more than 200 times as high as the pumping speed of the mechanical pump. The foreline pressure must thus rise to a pressure 200 times as high as the pressure at the entrance to the diffusion pump in order that the forepump can remove gas from the foreline at the same rate that the diffusion pump exhausts gas into the foreline. The pressure in the chamber rapidly decreases, thus decreasing the gas load so the high vacuum valve may be continuously (though carefully) opened without causing excessive foreline pressures. By the time the high vacuum valve has been fully opened, the chamber pressure has been reduced to a low enough value that the forepump can easily handle the gas load and the forepressure drops toward its normal equilibrium value. In a reasonably good vacuum system at equilibrium the foreline pressure will be about 1 mTorr.

In many vacuum systems the valves are remotely activated and will open fully without any provision for slow and careful operation. In such systems it is necessary to carry out the roughing procedure to a chamber pressure of less than 10 mTorr. In addition, the foreline will usually incorporate an expansion chamber to help limit the foreline pressure rise that occurs when the high vacuum valve opens.

We speak of these pressures in the range 1–1000 mTorr as if we always measure them precisely. It may be that with new thermocouple gauge tubes and controls we could be very careful in setting the zero point and could then measure the pressure very closely. Even with new equipment the effort is usually more trouble than it is worth, and with older equipment there is the added problem that the results are quite questionable. In practice, we set the zero point somewhere near but normally higher than the calibrated zero mark on the scale. We then become accustomed to what the normal readings may be for various stages of pumping and use these readings as our guide in the various stages of a vacuum procedure. This soon becomes second nature, and we find that not knowing the foreline pressure exactly is not really a difficulty.

Vacuum Chamber and Feedthroughs

Thus far we have not considered the vacuum chamber in any detail. It turns out that the vacuum chamber with vacuum feedthroughs may be the most complex part of a vacuum system. Historically, the most common vacuum chamber is composed of a 304 stainless steel baseplate about 1 inch thick and a glass bell jar of 18-inch diameter and 30-inch height. A Viton rubber boot gasket (L-shaped cross section) fitted onto the base of the bell jar provides a vacuum seal between the glass bell jar and the metal baseplate. The baseplate has a pumping port (usually 6 inch) to mate with the pumping system and, in addition, has a number of holes (usually 1 inch) through which various vacuum feedthroughs may be fitted. There is a trend to replace the glass bell jar with a stainless steel cyclinder having a top plate of about the same construction as the baseplate. This usually allows more size versatility and also allows more feedthroughs.

The most simple feedthrough, of the type depicted in Figure 6, serves merely to plug a feedthrough hole. We need not discuss dimensional details here, but we can see some basic principles. The plug covers the feedthrough hole, and the Viton O-ring provides a vacuum seal between the plug surface and the baseplate surface. The plug is contoured such that one contour fits the feedthrough hole and centers the plug, another contour contains the O-ring and seats against the baseplate to prevent excessive squeezing of the O-

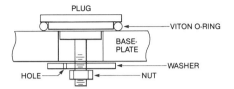

FIGURE 6   Blank plug for vacuum feedthrough hole.

ring, and the top contour makes the vacuum seal to the O-ring. The threaded rod is usually ¼-20. The nut is sketched as a hexagonal head, but a wingnut is usually more convenient. The function of the washer, obviously, is to bridge the gap. The hole in the washer is for access in leak detecting, which will be discussed in a later section.

The low voltage, medium current vacuum feedthrough depicted in Figure 7 is not greatly different from the blank plug of Figure 6. An insulated bushing serves to center the feedthrough and also to prevent excessive squeezing of the O-ring while providing electrical insulation between the feedthrough and the baseplate. An insulating washer provides electrical insulation between the nut and the baseplate. The metal washer provides protection for the insulating washer. This type of feedthrough is useful for voltages up to 50 V and currents up to 100 A. The insulating material is usually phenolic.

Currents in the range 100–500 A can be carried by a modified feedthrough, such as the one depicted in Figure 8a. The basic feedthrough is the same as Figure 7 except for a little additional length machined down to ⅜″ outer diameter (OD) at the threaded

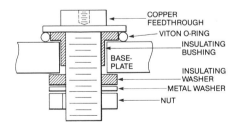

FIGURE 7   Low voltage, medium current feedthrough.

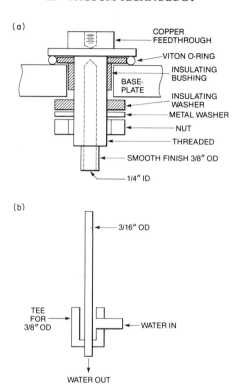

FIGURE 8 (a) Water-cooled feedthrough for low voltage, high current applications. (b) Tee arrangement.

end. The feedthrough is drilled out ¼″ inner diameter (ID) as indicated so that the tee-tubing attachment depicted in Fig. 8b may be attached. Water fed in and out as shown, via plastic tubing, can maintain the temperature of the feedthrough at acceptable levels for quite high levels of power dissipation inside the vacuum chamber. Tees and fittings to make such attachments are available from a number of companies and often are available in hardware stores.

A feedthrough of the type depicted in Figure 9 is suitable for either higher voltages or for radio-frequency (rf) power input into a vacuum chamber. The metallic feedthrough section, separated from the rest of the feedthrough, is shown on the left. One Viton

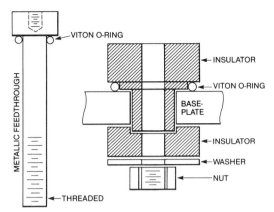

FIGURE 9 High voltage and RF power feedthrough.

O-ring provides the vacuum seal between the metallic feedthrough section and the top insulator. A second Viton O-ring provides the vacuum seal between the top insulator and the baseplate. The top insulator projects all the way through the baseplate and into an inset in the bottom insulator. Ceramic insulators of this type are available, but it is often more convenient to machine them from Teflon rod. This type of feedthrough is often used in situations in which high power levels are fed into the chamber and the feedthrough must then be water cooled. The modification of a low voltage feedthrough shown in Figure 8 is equally effective with the feedthrough depicted in Figure 9. Frequently this modification is extended to drilling through the entire length of the metallic feedthrough so that water can be brought to the object to which power is being fed. In this case the upper section of the metallic feedthrough is usually lengthened so that it may be threaded, usually ¼″ NPT. The modified feedthrough can handle power levels of up to several kilowatts.

The push–pull–rotary feedthrough depicted in Figure 10 allows the transmission of linear and rotary motion into the vacuum chamber. The rod that actually provides the motion has been omitted from this sketch. Two Viton O-rings provide the vacuum seal to the rod. The volume between these two O-rings is usually evacuated by connecting the pumping line to the foreline. The threaded

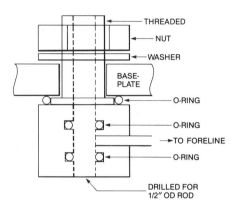

FIGURE 10   Push–pull–rotary feedthrough.

section is often grooved to allow evacuation of the air trapped below the washer and nut, but a cut from the ID to the OD of the washer is easier and just as effective. The washers in the power feedthroughs should also be cut from the ID to the OD to permit access for leak detecting.

There are a multitude of other vacuum feedthroughs, many of them substituting brazed metal-to-metal and ceramic-to-metal seals in place of O-ring seals. The basic principles are embodied in the five simple feedthroughs shown here, and there is little to be gained here by further illustrations of vacuum feedthroughs.

Vacuum valves are more complicated vacuum feedthroughs; nevertheless the same basic principles are employed. The mechanical aspects of vacuum valves usually are as simple as those of water valves, but the high vacuum valve is often somewhat more complicated mechanically. An understanding of vacuum seals together with some knowledge of mechanical devices enables one to disassemble and reassemble most vacuum valves.

Pumping Limits Due to Outgassing

In the section on pumping limits, cold traps, and baffles, we were left with the impression that backstreaming gases are the principal factor in limiting the ultimate vacuum which can be achieved. This is not the case. The principal factors in limiting the ultimate vacuum are found in the vacuum chamber. Not too surprisingly, one of these factors can be related to common experience. On a humid day, we commonly speak of the air as feeling heavy and damp, and many things, especially cloth items, feel damp even though there is no condensation in the form of fogging or dew. Conditions are not such that water vapor condenses out of the air, but many items feel damp. As we consider this fact we begin to realize that there must surely be a process other than condensation by which these items pick up water molecules out of the air. This process is known as adsorption. Adsorption occurs between all gases and all solid surfaces. A given gas will be adsorbed more strongly on some surfaces than on other surfaces. A given surface will adsorb some gases more strongly than other gases. With a given gas and a given surface, more gas will be adsorbed as a high pressure than at a low pressure. With a given gas, a given surface, and a given pressure, more gas will be adsorbed at a low temperature than at a high temperature.

Water vapor is a gas that is adsorbed very strongly and is most troublesome in vacuum systems. When a vacuum chamber is open and exposed to air, large quantities of water vapor are adsorbed onto the interior surfaces. When the chamber is closed and evacuated, the air is pumped out in a relatively short period of time. The main gas load to be removed by the pumping system is then gases, mostly water vapor, being desorbed from the interior surfaces of the vacuum chamber. This desorption of gases, normally referred to as outgassing, is responsible for most of the pumping time after the high vacuum valve has been opened. The increase in pumping time on humid days is very noticeable and measurable. Heating the chamber is very helpful in speeding up the desorption of gases, even in times of low humidity such as are experienced during the winter months. One is limited in the extent to which a chamber can be heated because of the use of O-rings. If the O-rings are lubricated with vacuum grease, as must be done with O-ring-sealed push–pull–rotary shafts, then heating is very strongly limited. The construction of Figure 10, leaving these O-rings external to the chamber, permits some cooling of the lubricated O-rings so that the other feedthroughs and the chamber may be brought to higher temperatures. Temperatures high enough to reduce pumping times to reasonable values can be achieved if the O-rings are made of Viton. O-rings made of other elastomers degrade too readily under even mild heating to be compatible with vacuum work.

Another highly effective method of speeding the outgassing is the use of a high voltage discharge during the pumping cycle. An ac voltage of about 5000 V applied to an electrode inside the vacuum chamber at pressures in the range 20–200 mTorr will maintain a discharge (or plasma) with a fairly high density of energetic ions. These ions impinge on various internal surfaces causing a high rate of desorption of adsorbed gases. This discharge cleaning is normally done during the roughing cycle.

A surprisingly effective method of speeding the outgassing is simply to maintain a flow of argon into the vacuum chamber while pumping. The argon flow should be at such a rate that the equilibrium pressure with the high vacuum valve open is between 10^{-5} and 10^{-3} Torr. We recall that we found that, at 1 mTorr of pressure, every atom or molecule on a solid surface is struck by a gas atom 267 times per second. Adsorbed molecules, being surface molecules, will therefore be struck approximately 3–300 times per second by argon atoms if the chamber pressure is 10^{-5} to 10^{-3} Torr. We have spoken of average velocities of gas molecules previously and found a handbook value of 4.85×10^4 cm/sec for the average velocity of an air molecule. The average velocity of an argon atom is 4.13×10^4 cm/sec. We have not spoken of the variation of velocities or the distribution of velocities of gas atoms. Velocities of gas atoms range from values much below to much above the so-called average velocity. Studying Figure 11 gives a simple idea of how and why the atoms within a collection of gas atoms should be expected to have a variety of velocities. Here, two identical gas atoms with identical speeds collide such that one of them (2) hits the other (1) broadside. We know that under these conditions all of the momentum and energy of 2 is transferred from 2 to 1 and that 1 also retains all of its original momentum and energy. Gas atom 2 is thus reduced to zero velocity, and gas atom 1 is increased in velocity to $v\sqrt{2}$. We can readily see that if a collection of gas atoms were to start with all atoms having equal speeds, collisions among them would soon introduce variations in velocities from near zero to velocities appreciably higher than the

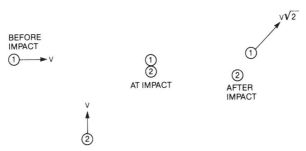

FIGURE 11 A collision between two identical gas atoms having initially identical speeds.

initial velocity. We can see that among the 267 gas atoms that strike an adsorbed molecule each second there must be some very energetic atoms. Some of these energetic atoms are certain to be of sufficient energy to strike the adsorbed molecule in such a way as to cause desorption. Heating generally seems to be the most effective means of speeding up outgassing, and high voltage discharge cleaning is next most effective. Nevertheless, the simple expedient of flowing argon gas through a system is surprisingly effective, and in many cases is the best method.

The flow of argon through a vacuum system is also beneficial with respect to reducing backstreaming problems. We have here spoken of a flow of argon high enough that the equilibrium chamber pressure is 1 mTorr, but a much lower flow rate would be almost equally effective. A flow rate resulting in an equilibrium chamber pressure of about 2×10^{-5} Torr is probably about the minimum flow rate that is effective in speeding outgassing and reducing backstreaming. Argon flow rates in this range of course will not reduce backstreaming from the diffusion pump to the chamber, but are quite effective in reducing backstreaming from the foreline to the diffusion pump. These flow rates increase the foreline pressure enough that backstreaming molecules in the foreline are caught in the argon flow and carried back toward the mechanical pump. The only backstreaming gas that can get to the diffusion pump and thence to the vacuum chamber is argon, which is inert and will not harm the vacuum process.

We have spoken of outgassing as if it consists only of gases that have been adsorbed by interior surfaces of the vacuum chamber while exposed to room air. There is also a component that is composed of gases that permeate the vacuum chamber through the bulk of the material composing the walls and seals of the chamber. Permeation through metal, glass, and vacuum-compatible ceramics is negligible in a demountable chamber using elastomer seals. Permeation through the elastomer seals can be quite severe, especially as the chamber temperature increases. We have mentioned that Viton is the elastomer best able to resist decomposition and therefore most compatible with vacuum work. Viton is also one of the elastomers having very low permeation rates and thus most compatible with vacuum applications. Nevertheless, higher temperatures do cause increased decomposition and permeation rates so it is preferable to work with the vacuum chamber at room temperature. If the chamber has been heated to speed up the outgassing, then the chamber should be cooled back down before starting the vacuum procedures.

Meissner Traps

Vacuum systems are frequently pumped down with no liquid nitrogen in the cold trap. If the ion gauge is turned on shortly after the high vacuum valve has been opened, the gauge reading will usually be in the low (10^{-5}) or high (10^{-6}) range. As liquid nitrogen is added to the cold trap, the ion gauge will suddenly begin to decrease and in 10–15 sec will drop by over an order of magnitude. The first impression is that the pumping speed has been greatly accelerated by the addition of liquid nitrogen to the cold trap. If we study the schematic representation of a vacuum system in Figure 5, we can see that this is not true. In actuality, before liquid nitrogen was added to the cold trap, the ion gauge tube was sensing gases that were fairly representative of the gas coming from the vacuum chamber. Most of this gas was water vapor. When liquid nitrogen was added to the cold trap, most of the water vapor was frozen out onto the liquid-nitrogen-cooled surface, and only the so-called noncondensable gases were able to reach the ion gauge tube. The ion gauge reading decreased by a very large factor, but the actual pressure in the vacuum chamber changed very little.

As we consider this representation of a vacuum system, we can see that the liquid-nitrogen cold trap should not be expected to have a great effect on the pumping speed. By the time the random motion of a water molecule has brought it to the vicinity of the cold trap, it is more likely that its subsequent random motion will bring it to the diffusion pump than back to the chamber. It makes very little difference in the pumping speed whether the water molecule is frozen out at the cold trap or left to meander along its random path, which will probably remove it from the system. Either way the water molecule is out of the system. We might gain a few percent in pumping speed by adding liquid nitrogen, but this makes very little difference in ultimate pumping time.

We do gain two advantages by adding liquid nitrogen to the cold trap. As discussed previously, this greatly reduces the backstreaming of condensable gases. This advantage may be important in some cases, but, as a rule, backstreaming in a good vacuum system is not a serious problem. The other advantage is that it freezes out the worst outgassing components so that the ion gauge tube senses only the so-called permanent or noncondensable gases. In a good vacuum system these gases are reduced to their ultimate minimum levels in a matter of minutes so the ion gauge reading will be low very shortly after the high vacuum valve has been opened because the permanent gases have been pumped out and the condensable gases are being frozen out. If there is an air leak into the chamber, there are large components of permanent gases (nitrogen and oxygen), and the ion gauge reading will be high. Reading the ion gauge while there is liquid nitrogen in the cold trap allows one to determine the integrity of a vacuum chamber within a matter of minutes. A low ion gauge reading means there are no leaks. It takes only a short time of working with a given vacuum system to enable one to know with certainty the ion gauge readings to be expected under various normal conditions. Departures from these readings are certain signs of malfunction.

It occurred to C. R. Meissner that a second liquid-nitrogen cold trap could be very effective in helping to remove water vapor and other condensable gases if it were within the vacuum chamber itself. Such traps, called Meissner traps, are often in the form of large coils of copper tubing or of large metal shrouds soldered to large coils of copper tubing. These are awkward in use and seldom better than the simple reentrant form illustrated in Figure 12. If this cold trap is made in the form of a 5-cm diameter cylinder 15 cm long, it has a pumping surface equal to the usual pumping port. On this basis, this form of Meissner trap should double the system pumping speed for water vapor. In actual practice, this form of Meissner trap will at least quadruple the system pumping speed for water vapor. If we study Figure 5 again, we can see that this is not unreasonable. In a chamber with a Meissner trap as in Figure 12, every water vapor molecule impinging on the cold trap will be frozen out onto the trap surface and immediately removed from the chamber. A water vapor molecule that exits through the pumping port in the baseplate has a high probability of striking a wall of the pumping tube or the high vacuum valve and bouncing back up into the chamber. The pumping speed of the pumping port is actually strongly restricted by the pumping tube and the high vacuum valve.

It may have occurred to some of us that a great deal of effort is being expended to remove water vapor from the vacuum chamber. We may wonder if this is necessary since water vapor is an oxide and therefore should be fairly inert. Many people with many years of experience have found that with atomically clean metal layers such as will be discussed in coming chapters, every type of gas molecule other than a noble gas is highly reactive. Water vapor is

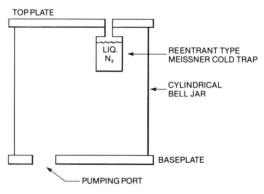

FIGURE 12 Cylindrical vacuum chamber with baseplate, top plate, and Meissner trap.

worse in this respect than other gases, but only because it is so much more difficult to remove water vapor than to remove other gases.

Leak Detecting

We have already broached the subject of air leaks into the vacuum chamber. It is a common belief that leaks in a vacuum chamber are not as serious as in a pressurized vessel. The reasoning that leads to this erroneous belief is that a vacuum chamber withstands a pressure differential of only 15 psi (pounds per square inch), whereas pressurized vessels must withstand pressure differentials of 3000 psi. To analyze this reasoning, we first convert these pressures to units with which we are familiar: 760 Torr and 152,000 Torr. It seems to be fairly reasonable to assume that a pressure change of 1% in a period of 10 years would be acceptable in either a vacuum chamber or a pressurized vessel. We can make a crude estimate of the area A of a leak of this size in a pressurized vessel of volume V by assuming that the air passes through this leak at the speed of sound v. The leak rate is 1% of 152,000 Torr from a volume V in 10 years (3×10^8 sec) or $L_p = (5.07 \times 10^{-6})$ V Torr/sec. The leak rate is also given by the product of the area, the pressure, and the velocity of sound (3.32×10^4 cm/sec) or $L_p = (5.05 \times 10^9)$ A Torr cm/sec. The volume must be in cubic centimeters and the area in square centimeters for these units to be consistent, in which case (5.05×10^9) $A = (5.07 \times 10^{-6})$ V or $A = (V \times 10^{-15})$/cm. A leak of this size in a vacuum chamber would result in an inflow of air at a rate equal to the product of the area ($A = V \times 10^{-15}$/cm), the pressure (760 Torr), and the velocity of sound (3.32×10^4 cm/sec); i.e., $L_V = (2.54 \times 10^{-8})$ V Torr/sec. We want to calculate the time t required for a leak of this size to change the pressure in our vacuum chamber by 1%. We assume that our vacuum chamber is 100 times the size of the pressurized vessel, i.e., of volume $100V$, and that the vacuum is 10^{-6} Torr, a fairly standard value. The expression for this leak rate is thus 1% of 10^{-6} Torr in a volume $100V$ in t sec or $L_V = V \times 10^{-6}/t$ Torr. Equating the two expressions for L_V, we obtain (2.54×10^{-8}) $V = V \times 10^{-6}/t$, or $t = 39.4$ sec. A leak rate that takes 10 years to affect a pressurized vessel will affect a vacuum chamber in less than a minute.

An air leak is the first matter of concern as a vacuum chamber is evacuated. With a given pumping system, one soon learns the normal time required to rough the system down to a pressure low enough that pumping can be valved over from the mechanical pump to the diffusion pump. If this time is exceeded, one is alerted to the probability of an air leak. If the air leak is so bad that the roughing cycle is unable to get the reading of the chamber thermocouple gauge (gauge 2 in Figure 5) below 200 mTorr, there is usually nothing that can be done other than readmitting air, disassembling the chamber, and inspecting for poor vacuum seals. One can be judicious and check first the seals that have been most recently altered.

If the air leak is at a low enough rate that the chamber can be roughed below 200 mTorr, one can often pinpoint the location of the leak before readmitting air to the chamber. If methanol (methyl alcohol) is applied to the point at which there is an air leak, the thermocouple gauge reading will usually change very quickly by a large amount. This change is usually an increase in pressure, but at times may be a decrease in pressure. Using a hypodermic syringe (and needle) of 10 cc size is a convenient method of squirting methanol on suspect vacuum seals. A syringe with finger- and thumb-rings is most convenient. Again one can be judicious and leak-chase first the feedthroughs and vacuum seals that have been most recently altered. One sees here the reason for using slotted washers on feedthroughs to provide access for leak chasing. Inserting the needle through the slot enhances the probability that the methanol will reach the O-ring and the potential leak. Once an air leak has been detected, we always waste our time trying to eliminate it by tightening nuts or bolts. After we realize that this is useless, we readmit air, disassemble the faulty seal, clean it, reassemble it, and evacuate the chamber.

The cause of an air leak is almost always lack of cleanliness. For example, a loose metallic flake, a piece of lint, or a hair is usually found between the O-ring and the sealing surface. Occasionally, the Viton O-ring is found to have been scratched, cut, or otherwise damaged. On rare occasions extreme carelessness has resulted in a scratch in the sealing surface. In these cases one must use metal sanding paper to grind out the scratch. One starts with 320-grit paper and polishes until the scratch is removed. The 320-grit sanding marks are polished out with 400-grit paper and final polishing is done with 600-grit. If 320-grit is unable to take out the scratch, the surface must be machined.

Most air leaks are at a rate so low that they are not detectable in the roughing cycle. Usually such a leak is found after the system has been valved over to the diffusion pump, the cold trap filled, and the ion gauge turned on. One then sees that the ion gauge reading is higher than it should be. Again the leak may be detected by application of methanol to the suspect vacuum seal. Now we observe the ion gauge reading instead of the thermocouple gauge reading. When methanol hits the air leak, the air leak becomes a methanol leak. The air is pumped out of the chamber in a matter of seconds, and the chamber is filled with methanol vapor. The methanol vapor approaching the ion gauge is frozen out on the cold trap, and the ion gauge reading goes down. Having detected the leak, we open up the chamber, correct the problem, and pump down again.

There are leaks that for one reason or another are not detectable by using methanol to probe suspect seals. Usually the reason is that the leak rate is so low that the sensitivity of the methanol method is not sufficient to give a detectable change in the ion gauge reading. In these cases one must have access to a helium leak detector, shown schematically in Figure 13. The heart of this system, the helium sensor, is some type of mass spectrometer or mass filter tuned to helium. This system has its own complete pumping system. The inlet is connected to the foreline of the vacuum system to be leak chased (such as in Figure 5), usually via rubber tubing connected to the foreline vent. Momentarily closing the foreline valve of the test system and opening the foreline vent valve of the test system will rough out the connecting tubing. The foreline valve of the test system is then reopened, and the throttle valve of the helium leak detector is cautiously opened while watching the ion gauge reading of the helium leak detector. There is a maximum allowable pressure at which the helium leak detector may be operated, and the throttle valve is opened until either this

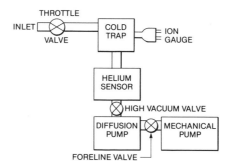

FIGURE 13 Schematic representation of a helium leak detector.

maximum ion gauge reading is reached or the throttle is fully open. The helium leak detector is thus sampling a portion of the exhaust of the diffusion pump. A small flow of helium from a pressure tank of helium is directed onto suspect vacuum seals. A hypodermic needle on the end of a length of rubber tubing connected to the helium tank pressure regulator provides a very convenient flow of helium. When the flow of helium is directed at the leaking seal, helium flows via the leak into the chamber, is pumped into the foreline, and is sampled and detected by the helium leak detector. This is the most sensitive method we have for detecting leaks. We can increase the sensitivity by interposing a valve at the intake of the mechanical vacuum pump of the test system and closing it so that the entire exhaust of the diffusion pump is sampled by the helium leak detector.

Maintenance of Vacuum Pumps

When a vacuum system fails to pump down properly, it is almost invariably because of a faulty seal in the vacuum chamber. On extremely rare occasions the fault may be found in either the diffusion pump or the mechanical pump. When the fault is found in a pump, it is almost invariably the result of a serious pumping accident and will not be mistaken for an air leak. Exercising reasonable care in the use and maintenance of a pumping system will prevent almost all problems of this nature.

The mechanical vacuum pump is, in principle, a very simple mechanism. Use and maintenance of these pumps are also quite simple. One routinely checks and adjusts the drive belt or drive system. One routinely oils the pump motor and maintains the oil level in the pump. The mechanical pump seldom requires more than this routine attention. It is important to recognize that the oil used in the vacuum pump is not the same as that used to oil the motor. Vacuum pump oils are special oils having vapor pressures that are relatively low compared to vapor pressures of motor oils. The motor oil need only lubricate. The vacuum pump oil must both lubricate and form vacuum seals inside the pump. There is enough oil seepage at seals that are directly exposed to the oil reservoir to provide lubricating-vacuum sealing oil films at the interior surfaces. Because of this seepage, it is not advisable to turn off a mechanical vacuum pump without venting it to atmospheric pressure. After venting to atmospheric pressure, the pump rotor should be turned a couple of revolutions to get the interior volumes up to atmospheric pressure also.

When a vacuum chamber is being roughed out, the mechanical pump is handling large quantities of gas. This gas is exhausted through the oil reservoir and picks up large quantities of oil in the form of oil mist or droplets. This can cause rapid depletion of the oil from the oil reservoir, and the mist can cause serious contamination of the laboratory as it deposits out of the atmosphere. To avoid these problems, we employ exhaust filters on the mechanical pumps to remove oil mist or entrained oil droplets from the exhaust gas and return the oil to the reservoir.

In the type of work in which we are engaged, the pump is not exposed to contaminants, except in the case of accident, and the oil will last for years except for routine replenishment to maintain oil levels. When it is necessary to change the oil, the pump should first be operated for 15 to 20 minutes in order to warm it (and the oil). In some extreme cases it may be necessary to drain the contaminated oil and add fresh oil before operating the pump to warm it. This costs a filling of oil, but is well worth it. After the warmed pump is shut off and drained, it should be turned on, and a small amount of oil should be poured into the inlet (at a reasonable rate) while the pump is running. If this oil comes out contaminated, the flushing should be repeated. If the second flushing oil comes out contaminated, a third flushing is called for. The amount of oil used in each flushing depends on the pump. In a small pump (up to 50 liters/min one might use 200 cc and in a medium-size pump (up to 500 liters/min) one might use 500 cc. The third flushing will have certainly cleaned out the interior pump surfaces, and, if the oil still comes out contaminated, the oil reservoir case will have to be removed. Paper towels are very effective for cleaning surfaces exposed by removal of the oil reservoir case. Solvents should never be used to clean any part of a mechanical vacuum pump. In replacing the oil reservoir case, we find it to be most convenient to use silicone rubber caulking instead of a gasket. Reasonable care will result in a clean, effective seal.

Use and maintenance of the diffusion pump are, if anything, more simple than use and maintenance of the mechanical pump. We are already aware that diffusion pump fluids are highly specialized oils much different from mechanical pump oil. Ordinarily there is no need to check or replenish this fluid. The diffusion pump becomes contaminated only by the same type of accident and at the same time as the mechanical pump. In the event of contamination, the contaminated oil is poured out, and the pump disassembled as much as possible. Soap, water, and elbow grease are then the most effective cleaning method. Thorough rinsing with hot water followed by thorough rinsing with methanol constitutes the final cleaning procedure. The pump is then reassembled, refilled with diffusion pump fluid, and replaced in the pumping system.

Other Vacuum Pumps and Gauges

One might get the impression that there are only two types of vacuum pump, namely, the two we have discussed. Although these two types of pumps are employed for the majority of vacuum pumping applications, there are a great number of other types. There are other designs of mechanical pump for the same applications as the oil-sealed rotary pump described here. There are mechanical pumps designed to augment or replace diffusion pumps. There are many designs of diffusion pumps. There are sorption pumps designed to perform the mechanical pump function. There are cryogenic pumps and ion pumps designed to replace diffusion pumps. Some of these pumps function in place of ion gauges. There are many vacuum gauge designs other than the two types described here. These pumps and gauges have not been discussed in this book because they are somewhat separated in scope from the intention of this book. It is felt that the reader, upon completion of this book, will be prepared to explore other literature to learn the nature and applications of these other pumps and gauges.

CHAPTER III

VACUUM EVAPORATION

At this point it is hoped that our orientation in the field of vacuum technology is such that we are prepared to consider vacuum evaporation of metals. It is expected that we understand that a metal vapor is a gas just as air is a gas. It is expected that we understand that Avogadro's law is as true for a metal vapor as it is for water vapor or oxygen or any other gas. In principle, Loschmidt's number is the same for a metal vapor or for water vapor or for any other gas. In practice, gases such as metal vapor or water vapor or gasoline fumes do not achieve standard pressure at standard temperature because their vapor pressures at standard temperature are too low. Therefore Loschmidt's number exists only in principle for these gases.

Vapor Pressure

We have spoken of vapor pressure several times without being precise about the meaning of vapor pressure. To define vapor pressure, we assume that we have a material within a chamber that has been pumped to a negligibly low pressure of air and sealed. We have a means of holding this chamber (and contents) at any temperature we choose and a means of measuring the gas pressure within the chamber. The only gas within the chamber is the vapor of the material within the chamber. We assume that the chamber, the seal, and the pressure sensor are made of material having negligible vapor pressure at temperatures involved in this hypothetical experiment. The material in the chamber can be almost anything: it might be argon, it might be mercury, it might be aluminum. We start out with the chamber at a temperature low enough that the gas pressure (or vapor pressure) is too low to register on our pressure sensor. We find it natural to refer to the pressure as gas pressure if the material is argon, but it seems more natural to say vapor pressure if the material is mercury or aluminum. We bring the chamber up to a temperature at which the gas pressure will be high enough to register on our pressure sensor and the atoms will begin to sublime (or evaporate) from the solid (or liquid) surface of the material in the chamber. The gas pressure in the chamber rises to an equilibrium value that is characteristic of the material in the chamber at that temperature. At this equilibrium gas pressure, gas atoms are impinging and condensing on the material surface at exactly the same rate as atoms are evaporating from the material surface. The resulting equilibrium gas pressure is the vapor pressure of that material at that temperature.

Evaporation Temperature and Pressure

We recall that gas pressure is proportional to the numerical density of the gas; i.e., to the number of gas atoms per cubic centimeter, which we have denoted by the letter n. We also recall that gas atoms are constantly in motion with a distribution of velocities ranging from zero to very high values and that there is an average velocity, which we have denoted by the letter v. Finally, we recall that all surfaces within the chamber in our hypothetical experiment are bombarded by gas atoms at a rate $\frac{1}{4}nv$. In particular, the surface of the material is bombarded by gas atoms at the rate $\frac{1}{4}nv$. The impinging gas atom, being the same as the surface atoms, will condense on the surface. At equilibrium, there must be as many atoms leaving the surface as there are impinging on the surface. This is an important concept that bears repeating in slightly different wording. At equilibrium in an experiment such as we have here hypothesized, atoms of the material are being evaporated from the material surface at exactly the same rate as gaseous atoms of the material are impinging against the material surface. This rate we know to be $\frac{1}{4}nv$. Thus we have that atoms are ejected from a surface at a rate $\frac{1}{4}nv$, where n is the number of gas atoms per cubic centimeter under equilibrium conditions of the material at the existing temperature. This is a very important result; it gives us an expression, $\frac{1}{4}nv$, for the rate at which material may be evaporated. We now seek to convert this expression to measurable quantities.

We recall Loschmidt's number, 2.69×10^{19} gas atoms/cm^3, which is the value of n for all gases at normal temperature (0°C) and pressure (760 Torr). We know that n is directly proportional to the pressure p, but we have not considered the effect of the temperature T. We are quite aware that pressurized dispenser cans may build up to explosive pressure if exposed to heat. We know that letting some of the gas out of an over-pressured can will decrease the pressure. We know that this results in decreased n. We thus know that higher temperature results in lower n, and lower temperature results in higher n if pressure is to remain at a constant value. We are not greatly surprised to learn that n is inversely proportional to T. We thus have $n = kp/T$, where k is the constant of proportionality. We note here that the temperature T is measured on the absolute temperature scale known as the Kelvin scale. A temperature on the Celsius scale is converted to that temperature on the Kelvin scale simply by adding 273 to the Celsius temperature. Normal temperature is 0°C or 273°K. Substituting into $n = kp/T$ ($n = 2.69 \times 10^{19}$, $p = 760$, $T = 273$), we find $k = 9.66 \times 10^{18}$. Thus $n = 9.66 \times 10^{18}\, p/T$ gives the number of gas atoms per cubic centimeter at a pressure p (Torr) and temperature T (°K).

In the description of the operation of the thermocouple vacuum gauge we said that gas atoms that impinge against the heater wire acquire higher velocities consistent with the wire temperature. We had assumed a general familiarity with the fact that velocities of gas atoms are higher at higher temperatures and lower at lower temperatures. The exact relationship is that the average velocity of gas atoms is directly proportional to the square root of the temperature. In the discussion of average velocities of gas atoms and mean free paths we found handbook values of 4.85×10^4 cm/sec for an air molecule and 6.15×10^4 cm/sec for a water molecule. We probably should then have pointed out that average velocities of gas atoms are higher for lighter atoms and lower for heavier atoms. The exact relationship is that the average velocity of gas atoms (molecules) is inversely proportional to the square root of the atomic (molecular) weight. We thus have $v = b(T/M)^{1/2}$. Knowing that $v = 6.15 \times 10^4$ cm/sec for a water molecule of molecular weight $M = 18$ at a temperature of $T = 273°K$, we find that $b = 1.58 \times 10^4$. Thus $v = (1.58 \times 10^4) (T/M)^{1/2}$ gives the average velocity of a gas atom of atomic weight M when the gas is at a temperature T.

Substituting these expressions for n and v into $\frac{1}{4}nv$ gives $(3.82 \times 10^{22}) p(TM)^{-1/2}$ as the number of gas atoms of atomic weight M impinging per square centimeter each second on a surface when the gas pressure is p and the temperature is T. We can measure p and T, and, as long as we know the type of gas we are using, we can find the value of M in a handbook. We recall that we earlier calculated that surfaces in a chamber at a pressure of 1 mTorr would be bombarded by air molecules at a rate of $4.28 \times 10^{17}/cm^2$ sec. We can check this result by substituting $p = 10^{-3}$ Torr, $T = 273°K$, and $M = 29$ into our new expression and find that the rate of impingement is 4.29×10^{17} air molecules/cm^2 sec. The agreement at least gives us the feeling that we are being consistent.

The original importance of the expression $\frac{1}{4}nv$, or, in our new form, $(3.82 \times 10^{22}) p(TM)^{-1/2}$, was to see that surfaces in a chamber were bombarded by gas atoms at a high rate even at very low gas pressures. The added importance of this expression now is that it gives the rate at which atoms of a material of atomic weight M leave the surface of the material when the temperature of the material is T. The pressure p in this expression in this case refers to the equilibrium vapor pressure of the material at the temperature T. We have now a means of calculating the rate at which atoms are evaporated from a surface. We shall find that, even though we shall not use this in actual applications, this expression will help to give us some insight into the evaporation process.

We now want to narrow the range of materials we shall consider to those elements that will be used in coating parts or substrates by vacuum evaporation. These materials are listed in Table I. These materials are in the solid state under normal conditions. To coat parts or substrates with one of the materials, we use, as a source, a suitable piece of the material placed in or on a heater. We position parts or substrates somewhat near (10 to 30 cm) the source, with the side to be coated facing the source. The vacuum chamber is closed and evacuated and power is applied to the heater. When the material reaches a suitable temperature, atoms begin to evaporate (or sublime) from the source. These atoms leave the source in

TABLE I[a]

Element	Symbol	M	d	MP (°K)	BP (°K)	Ev T (°K)	Ev p (Torr)
Aluminum	Al	27.0	2.70	930	2770	1460	7.9
Chromium	Cr	52.0	7.20	3700	2700	1700	16.3
Copper	Cu	63.5	8.92	1356	2850	1580	17.6
Gold	Au	197.0	19.3	1336	3120	1670	22.2
Germanium	Ge	72.6	5.35	1210	2850	1400	9.3
Iron	Fe	55.8	7.86	1810	3100	1770	17.5
Magnesium	Mg	24.3	1.74	920	1370	680	3.6
Molybdenum	Mo	95.9	10.2	2890	4200	3000	22.5
Nickel	Ni	58.7	8.90	1730	3100	1810	19.5
Niobium	Nb	92.9	8.57	2740	4200	—	—
Palladium	Pd	106.4	12.0	1830	3200	1810	19.6
Platinum	Pt	195.1	21.45	2040	4100	2460	30.1
Silicon	Si	28.1	2.33	1680	2800	1800	7.4
Silver	Ag	107.9	10.5	1230	2470	1330	14.6
Tantalum	Ta	180.9	16.6	3250	4200	3470	28.7
Titanium	Ti	47.9	4.50	1940	3200	2000	11.5
Tungsten	W	183.8	19.35	3660	5000	3720	34.4
Vanadium	V	50.9	5.96	2160	3200	2160	15.3
Zirconium	Zr	91.2	6.49	2120	3200	2710	14.0
Quartz	SiO_2	60.1	2.65	1970	2200	—	—
Pyrex	—	—	2.23	1250	—	—	—
Stainless steel	—	—	8.0	1700	—	—	—

[a] M, molecular weight; d, density; MP, melting point; BP, boiling point; Ev T (Ev p), temperature (vapor pressure) at which the deposition rate at 10 cm from a 1-cm^2 source will be 25 Å/sec.

straight line paths and, assuming a good vacuum, continue in these straight line paths until they impinge on a solid surface (either parts, substrates, fixturing, or chamber walls). These solid surfaces are at temperatures much below the source temperature, and the impinging atoms give up their excess energy to the solid surface. The impinging atoms thus acquire a temperature at which their vapor pressure is so low as to be negligible, and therefore they cannot remain in the gaseous state so they must adhere to the solid surface.

In essence, we had defined the vapor pressure of a material as that pressure at which the gaseous phase is in equilibrium with the liquid (or solid) phase of the material when both phases are at the same temperature. Under equilibrium conditions gas atoms are depositing on the surface of the material at the rate (3.82×10^{22}) $p(TM)^{-1/2}$, and surface atoms are evaporating (or subliming) from the material at the rate $(3.82 \times 10^{22})\ p(TM)^{-1/2}$, where M is the atomic weight of the material and p is its vapor pressure at the temperature T. It is clear that the entire chamber must be at temperature T if both phases of the material are to be at temperature T. Our conditions now are much different so equilibrium is not maintained. The material is now at temperature T (a high temperature) so surface atoms are still evaporating (or subliming) at the rate $(3.82 \times 10^{22})\ p(TM)^{-1/2}$, but the chamber walls and other interior surfaces are at room temperature (a very low temperature relative to T). The evaporated atoms, after leaving the surface of the material, impinge against surfaces which are at temperatures so low that the vapor pressure of the material is not detectable. These evaporated atoms therefore adhere to these surfaces and do not return to the gas phase. The only material atoms in the gas phase are those traveling away from the material surface so no atoms are depositing on the material surface. As a consequence, the surface loses atoms at the rate $(3.82 \times 10^{22})\ p(TM)^{-1/2}$, and the cooler surfaces to which the evaporated atoms adhere become coated with a layer of this material of thickness dependent on the rate of evaporation, the time of evaporation, and the distance of the surface from the source of evaporating atoms.

We can make an interesting use of the expression (3.82×10^{22}) $p(TM)^{1/2}$, the number of atoms evaporated per square centimeter each second from a material of molecular weight M and having a vapor pressure p at temperature T. If the exposed surface area from which atoms can be evaporated is A and the material is maintained for a time t at temperature T, then the total number of atoms evaporated will be $N = (3.82 \times 10^{22}) pAt(TM)^{-1/2}$. We know that the weight of a single atom is given by $(1.66 \times 10^{-24}) M$ so the total weight of material evaporated is given by $m = (1.66 \times 10^{-24}) MN = 0.06434pAt(M/T)^{1/2}$.

To make use of this expression, we want now to introduce something that is somewhat outside the rest of the content of this book. We want to braze together, face to face, a stack of 21 copper sheets. Each sheet is 1.27 cm $\times$ 1.27 cm $\times$ 0.01 cm. We want to use a copper–silver brazing alloy, and we want the interfaces to be as free of contaminants as possible. If we use a braze foil at each interface, we have four faces that can introduce contaminants. The two braze faces are not very reactive and therefore do not introduce much contamination, but copper is much more reactive than we ordinarily realize, and the two copper faces introduce an appreciable amount of contamination. We bypass this problem by using techniques that are discussed later. We make these parts atomically clean by a process called back sputtering or sputter cleaning and immediately coat them with a (5×10^{-4})-cm-thick layer of silver. Each face now has within its surface the copper and the silver needed for the braze. When these are exposed to air, the copper surface is protected by the silver coating and is not contaminated. The silver, being much less reactive than either the copper or the braze alloy, introduces a minimum of contamination. The braze alloy now becomes superfluous and is not used because each surface of the now silver-coated copper has the braze components within itself. We have thus reduced the number of surfaces from four to two, both of which are cleaner than the original four. We now stack the 21 coated copper pieces together with a ceramic spacer under the stack, a ceramic spacer on top of the stack, and a weight to keep the copper pieces pressed together. This is placed in a vacuum furnace, brought up to a temperature just below the brazing temperature (1075°K), held there for 10 min, brought up to brazing temperature (1075°K), held there for 5 min, and then allowed to cool.

It is quite natural to wonder how much silver has evaporated or sublimed during this brazing cycle. As a first approximation, we assume that the parts have been held at full temperature (1075°K) for 20 min. We therefore have $T = 1075°K$ and $t = 1200$ sec. We find vapor pressure curves that give the vapor pressure of silver at 1075°K as $p = 5 \times 10^{-5}$ Torr. We find in Table I that the molecular weight M of silver is 107.9, and the only additional factor we need is the exposed area of silver. The silver is exposed only at the edges of the 20 interfaces. With 4 edges at each interface, there are a total of 80 edges. Each edge is 1.27 cm long and has an exposed silver thickness of $2 \times 5 \times 10^{-4}$ cm, or 10^{-3} cm, so the area of silver exposed at each edge is 1.27×10^{-3} cm². The total area exposed for evaporation is thus $A = 80 \times 1.27 \times 10^{-3}$ cm², or $A = 0.1016$ cm². We can now calculate that the total amount of silver evaporated during the brazing cycle is $m = 1.224 \times 10^{-4}$ g. Since the density of silver is 10.5 g/cm³, the total volume of silver evaporated is 1.166×10^{-5} cm³. Since the total exposed area is 0.1016 cm², the depth of silver lost from each edge is 1.148×10^{-4} cm, or 0.01%. This is quite acceptable, so we conclude that our brazing operation will not be hampered by the amount of silver lost through evaporation. We now want to return to consideration of vacuum evaporation for the deposition of thin film coatings onto parts and substrates.

We discuss the difficulties of trying to calculate coating thicknesses and coating uniformity from the geometry and basic principles later. We shall find that it will be necessary to go more by trial and error than by calculation. At this point we use an example that is artificial, but is not too far from a realizable situation. We assume that in a vacuum chamber we have a material of surface area 1 cm² from which atoms are being evaporated. We assume that these evaporated atoms are being deposited onto a 100-cm² surface that is positioned about 10 cm away from the source. We assume that none of these evaporated atoms deposits anywhere else. We want to deposit a 2500-Å (angstrom) coating over the 100-cm² surface in 100 sec.

This is our first introduction of the angstrom. The angstrom, a unit of length equal to 10^{-8} cm, is in widespread use. There is some effort to introduce the nanometer (nm), a billionth of a meter, but use of the nanometer is minimal. Another widely used unit besides the angstrom is the micron. We remember that the micron was introduced in pressure measurement as being a term equivalent to a millitorr. It has been noted that the micron had originated as being a millionth of a meter. It is still a millionth of a meter, and it is fairly common term in specifying thickness of coatings. A micron is 10^4 Å, or 10^{-4} cm. Another fairly common term in specifying thickness of coatings is the microinch. A microinch is equal to 254 Å, but, for most calculations, where errors up to 2% are usually acceptable, we take a microinch as being 250 Å.

From the coating thickness (2500 Å, or 2.5×10^{-5} cm) and the coating area (100 cm²) we calculate the coating volume as 2.5×10^{-3} cm³. We recall that to calculate the size of an atom we first found its weight by multiplying its atomic weight M by the mass of a nucleon (1.66×10^{-24} g). We then divided the weight of one atom by the density d of the material in solid form to find the volume occupied by one atom, (1.66×10^{-24}) M/d. If we now divide the volume of the desired coating by the volume occupied by one atom, we have the number of atoms deposited: $(2.5 \times 10^{-3}$ cm³) $[(1.66 \times 10^{-24})$ $M/d] = (1.51 \times 10^{21})$ d/M. Since these (1.51×10^{21}) d/M atoms have all been evaporated from the 1-cm² source, the source has emitted (1.51×10^{21}) d/M atoms/cm². Since these atoms were evaporated in a time of 100 sec, the evaporation rate is (1.51×10^{19}) d/M atoms/cm² sec. We set this equal to our expression for the evaporation rate (3.82×10^{22}) $p(TM)^{-1/2}$ and solve for p to get $p = (3.95 \times 10^{-4})$ $d(T/M)^{1/2}$. We must find values of p and T (on vapor pressure curves) that are consistent with this equality.

If we wish to deposit magnesium, we find in Table I that $M = 24.3$ and $d = 1.74$ so $p = (1.39 \times 10^{-4})$ $T^{1/2}$. We find vapor pressure curves similar to those of Figure 1 of Chapter 2 only easier to read, make the guess that $p = 10^{-2}$ Torr will provide the desired coating rate, find that $T = 730°K$ will give $p = 10^{-2}$ Torr, and calculate $p = (1.39 \times 10^{-4})$ $(730)^{1/2} = 3.70 \times 10^{-3}$ Torr. This is not 10^{-2} Torr so we do not have an equality, and therefore we do not have the solution. We then make the guess that $p = 3.6 \times 10^{-3}$ Torr, which can be achieved at $T = 680°K$, will give an equality and calculate $p = (1.39 \times 10^{-4})$ $(680)^{1/2} = 3.64 \times 10^{-3}$ Torr. This is close enough to 3.6×10^{-3} Torr that we can conclude that our evaporation source

operated at about 680°K will result in an evaporation rate close to
the desired 2500 Å in 100 sec, or 25 Å/sec. In this way we have
obtained the values listed in Table I under Ev T and Ev p.

If we wish to deposit aluminum, we have $M = 27.0$ and $d = 2.70$,
so $p = (2.05 \times 10^{-4})\, T^{1/2}$. Again we make the guess that $p = 10^{-2}$,
find $T = 1490°K$, and calculate $p = 7.92 \times 10^{-3}$. We then make the
guess that $p = 7.8 \times 10^{-3}$, find $T = 1460°K$, calculate $p = 7.84 \times
10^{-3}$, and conclude that 1460°K is the desired source temperature.

If we wish to deposit platinum, we have $M = 195$, $d = 21.45$, and
thus $p = (6.07 \times 10^{-4})\, T^{1/2}$. We guess $P = 10^{-2}$, find $T = 2360°K$,
and calculate $p = 2.95 \times 10^{-2}$. We then guess $p = 3.0 \times 10^{-2}$, find
$T = 2460°K$, calculate $p = 3.01 \times 10^{-2}$, and conclude that 2460°K is
the desired source temperature.

These results are listed in Table I, and we have calculated them
here only to see the effect of source temperature on vapor pressure
and hence on the rate of evaporation. We see that a 1% change in
source temperature results in an approximately 15–20% change in
evaporation rate. As a consequence, control of evaporation rate or
deposition rate is somewhat difficult but well within our capability.
It is interesting that the range of vapor pressures required for the
selected coating geometry and rate covers an order of magnitude
over the variety of elements listed, but this poses no particular
problem. The upper end of the temperature range here is somewhat high so heat losses due to radiation alone from even a small
evaporation source can exceed a kilowatt and heating of fixtures
and substrates can introduce difficulties.

We may at times want to deposit coatings at lower rates, in
which case we will operate the source at lower temperatures. More
frequently, we will want higher rates so we will operate at higher
temperatures. In the case of aluminum, for instance, we have a
nominal source temperature of 1460°K. To get a deposition rate of
2.5 Å/sec, we would need a source temperature of 1340°K. At 250
Å/sec the source temperature would be 1610°K. We see that an
order-of-magnitude change in evaporation (or deposition) rate can
be achieved by a relatively small change in source temperature.

We note that the nominal evaporating temperature for chro-
mium and for magnesium is appreciably below the melting point.
These materials are therefore deposited by sublimation from the

solid state. A number of materials, such as titanium, have a melting point so close to the evaporating temperature that it is often advantageous to settle for a lower coating rate so that the source material can be kept in the solid state during the coating operation. Many of the materials listed in Table I must be heated well beyond the melting point before the vapor pressure becomes high enough to provide a useful rate of evaporation.

It is a not uncommon misconception that a higher vacuum may result in an increased rate of evaporation. As we review our analysis of evaporation, we see that there are no grounds for this. The rate of evaporation is independent of the degree of vacuum as long as the vacuum is high enough that there are not enough residual gas atoms present to impede the escape of atoms from the source surface. We will recall that a vacuum of 1 mTorr or higher (a pressure of 1 mTorr or lower) meets this requirement. The degree of vacuum, as long as it is 10^{-3} Torr or higher, therefore has absolutely no effect on evaporation rate.

Vacuum Evaporation Sources

A schematic diagram of an often used and very simple vacuum evaporation source is shown in Figure 1. The sketch shows the tungsten wire as a single-strand wire, but it is most commonly a three-strand braided wire with each strand having a diameter of 0.05 to 0.1 cm. The source length is about 10 cm with a V bend having 1-cm arms at the center. The variable transformer is most commonly 0–220-V ac operating from a 220-V ac supply line. The step-down transformer usually has optional taps of 5 V, 10 V, and 20 V with the 5-V option chosen for this source. As the output from the variable transformer is turned up from zero, the evaporation source is resistance heated by the electrical current through it. This is exactly the same way that the filament of an electric light bulb is heated.

The material to be deposited (referred to as the evaporant) is in the form of one or more U-shaped staples (referred to collectively as a charge) hung at the point of the V in the wire. Successful use of this source requires that the evaporant be able to wet the tungsten

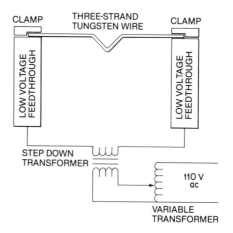

FIGURE 1 Loop filament evaporation source.

wire so as to be able to cling to the wire after melting. The disadvantage of this is that the evaporants having this characteristic
have also the characteristic of alloying with tungsten. There is then
a strong probability that some tungsten will be in the deposited
film, especially toward the end of the charge. Replenishing the
charge to alleviate this problem is usually not successful because
the alloying reduces the diameter of the wire a great deal, and the
life of the wire is very short in terms of both coating time and
repeated use. One does not usually attempt coatings of more than
1500-Å thickness with this type of source. Nevertheless, this
source will provide acceptable coatings of many types of material,
including Al, Cu, Fe, Ni, Au, Pd, and Pt.

This source is essentially a point source, and it deposits a fairly
uniform coating over a spherical surface centered at the source.
Positioning parts equidistant from the tip of the V results in fairly
uniformly coated parts. As one becomes experienced with a particular setup with this type of source, one may find that parts positioned in some directions from the source get heavier coatings than
the norm and parts in other directions get lighter coatings than the
norm. Increasing the distance from the source will decrease the
coating thickness, and decreasing the distance will increase the

FIGURE 2 Multiple-loop evaporation source.

coating thickness so uniformity can be improved easily. The question of getting the proper coating thickness is discussed in a later section.

Somewhat heavier coatings can be deposited from sources of the type sketched in Figure 2. Only the wire is shown in this sketch. Again the wire is shown as a single strand, but it is most commonly a three-strand braided wire. The source shown has 4 loops, but as many as 15 loops have been used. It is much more convenient to wind these as helical coils because the coiled form is stronger and more compact than a succession of V's would be. This source in operation is a multiplicity of point sources and more trial and error is required to find where parts must be positioned to achieve uniform coatings. The source length is usually about 10 cm with ~1-cm-diameter coils wound at a pitch such that the coiled section is about 4 cm long. This source, like all other sources mentioned herein, is commercially available at very reasonable cost from a number of suppliers. These suppliers also provide literature with a great deal of technical data on their evaporation sources and on materials available for evaporation. This literature is a valuable source of suggestions as to techniques and precautions to be used in evaporating each type of material.

The wire basket source shown in Figure 3 is quite versatile. Again the wire is most commonly a three-strand braided wire. The charge for this source can be in any form or shape as long as it is small enough to fit into the basket and large enough that it will not slip out through the spacings between coils. If the charge melts upon being heated, it need not wet the wire because it will coalesce

FIGURE 3 Wire basket source.

into a blob that will not flow out through the spacings. Since this source will retain solid metals, nonwetting liquid metals, and wetting liquid metals, it can be used for a greater variety of materials than the more simple sources of Figure 1 and 2. These three types of wire sources have in common that they are almost invariably made of braided wire for strength and flexibility. They are also similar in size, being about 10 cm long and of the order of 1 cm in vertical dimension.

The dimpled boat source depicted in Figure 4 has the advantage that it can hold a bigger charge than the wire sources and can therefore be used to deposit thicker coatings. It has the disadvantage that evaporated atoms leave only in the upward direction so parts can be positioned only above the source. Parts are positioned essentially in a hemispherical pattern above the source and improved uniformity of coating is achieved by trial and error. The dimple in the boat shown in Figure 4 has a circular shape, but an elongated shape is more common. The dimple is about 0.3 cm deep, and the overall dimensions of the boat are about 10 cm long and 1 or 2 cm wide. Wire sources are normally tungsten, but boat sources are made of molybdenum, tantalum, or tungsten. Sheet stock from which these sources are fabricated is usually 0.01–0.02 cm thick. Power requirements for these sources are appreciably higher than for wire sources.

Open sources similar to the wire and boat sources described here are commercially available at very reasonable cost from a number of suppliers of vacuum equipment. One may select from a profusion of design varieties stocked for immediate shipment. Also

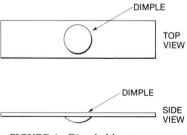

FIGURE 4   Dimpled boat source.

TOP VIEW

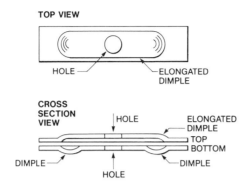

HOLE — ⌐ELONGATED DIMPLE

CROSS SECTION VIEW

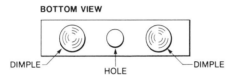

HOLE   ELONGATED DIMPLE   TOP   BOTTOM

DIMPLE   DIMPLE   HOLE

BOTTOM VIEW

DIMPLE   HOLE   DIMPLE

FIGURE 5   Simple furnace or covered boat source.

stocked are a variety of closed sources, similar in principle to that shown in Figure 5. Material to be evaporated is placed in one or both dimples in the bottom section. The two sections are clamped to the power feedthroughs, making a simple furnace. As the source heats up, the little chamber formed between the top and bottom sections becomes filled with evaporant vapor that streams out of the holes. In practice these sources are made with only one hole, and the hole is usually in the top section. The design of even the most simple source of this type is a bit more sophisticated than the sketch in Figure 5 since it has extended side walls to guard against warpage problems. The closed source design was developed to prevent evaporant specks from being ejected and striking the substrates. This is often a problem with open sources, especially open boat sources, and most especially with materials that do not wet the source surface.

With furnace-type enclosed sources one tends to feel that the evaporated atoms will leave the source in a preferred direction perpendicular to the plane of the opening. This is the case, and positioning substrates for coating is strictly a trial-and-error matter. Knudsen's cosine law for the effusion of gases from an isothermal enclosure is often cited as a starting point for a first estimate. Knudsen's cosine law gives $\frac{1}{4}nv\pi^{-1}\cos\theta$ as the rate at which atoms stream from the opening in a direction at an angle θ from the perpendicular. We recognize the $\frac{1}{4}nv$ portion of this expression as the expression we derived for the rate at which atoms impinge on a surface. The $\pi^{-1}\cos\theta$ portion is just the fraction going in the θ direction. This expression is correct only when the opening is small enough that the equilibrium between the vapor phase and the evaporant is not disturbed by the loss through the opening. In a practical source, the opening has to be large to get an acceptable coating rate so it is not possible to maintain equilibrium. Thus Knudson's cosine law is not really applicable to common enclosed sources.

Knudsen's cosine law is valid to a very close approximation for open sources. Theoretical justification for this is somewhat beyond the scope of this book, but we can see some of the reasoning involved. We recall that vapor pressure is defined as the pressure at which the rate of evaporation from a surface is equal to the rate at which vapor-phase atoms impinge on the surface. This equilibrium condition requires that the distributions of velocities and of directions of evaporating atoms must not differ from those of condensing atoms. Therefore atoms must be evaporating from the surface with exactly the same distribution as those with which they would escape from a small opening in a chamber which is at equilibrium.

In evaluating the coating thickness distribution from a source, the simple source of Figure 1 can be shown to approximate a point source closely as we previously assumed. Masking or shading by sections of the wire and filament supports complicates the distribution of coating in some directions. The multiple-loop source is merely the sum of several point sources, but directional and additional masking factors add further complications. Evaluation of open boat sources requires integrating point source distributions over the surface of the boat, but the shape of the boat complicates this. All sources suffer complications arising from warpage and from the variable distribution of evaporant material over the source surface. As a consequence, one is left with the need to determine the best arrangement for achieving uniform coating distributions by trial and error.

At one time or another we are certain to encounter one or another of three problems that occur with these resistance-heated sources. When there is alloying between the evaporant and the source material, there is contamination of the deposit. Where high purity is required, this can be a problem. There are practical limits to the amount of power that can be applied, and this can be a problem where extremely high rates or extremely high temperatures are required. The material of which the filament or boat is fabricated must have a negligible vapor pressure at the evaporation temperature of the evaporant. How can the high temperature materials, such as molybdenum, tantalum, and tungsten, be deposited? Of what material can the source be fabricated?

Electron-Beam Sources

These problems can be circumvented to a great extent by using electron bombardment heating rather than resistive heating of the evaporant. Using resistive heating, we first heat the source by passing an electric current through the source. The source in turn heats the evaporant by radiative heat. If, after melting, the evaporant wets the source, the evaporant is then heated by conduction. In either event, the source is heated first by the applied power, and the heat is then passed on to the evaporant. In electron bombardment heating, the evaporant may be directly heated. A simple electron-beam (E-beam) system is depicted in Figure 6. The loop filament is larger and is heated to a higher temperature than the filament of an ion gauge because greater electron emission is needed. The high voltage power supply provides 5000–10,000 V to accelerate the electrons from the filament to the evaporant. The electrons dissipate their acquired energy to the evaporant upon impact against the evaporant surface. Thus the evaporant is heated directly and energy utilization is much more efficient than with resistance-heated sources. Power input with an E-beam source can

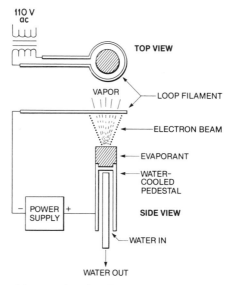

FIGURE 6   Simple E-beam evaporation source.

be much greater than with a resistance-heated source if the E-beam system is properly designed.

Since the support structure for the evaporant is water cooled, the evaporant does not wet or alloy with the source and remains pure. This eliminates the problem of the filament, boat, or other support structure introducing impurities into the coating. The E-beam system also resolves the problem of evaporating high temperature materials. There are not any support structure problems so the boat, hearth, or other support structure can be fabricated of almost any material. Greater amounts of power can be supplied, and this power is utilized with more efficiency so much higher temperatures can be reached.

The most severe problems that are encountered with E-beam systems are related to arcing and electrical discharges that result from the high voltage in the presence of high vapor densities. Some impurity problems may arise because vapor may become ionized, resulting in interactions with the hot filament that lead to impurities in the deposited film.

Evaporation of Alloys and Compounds

Up to this point our discussion has centered on the evaporation of metallic elements. We have not considered the evaporation of alloys or compounds. In looking at vapor pressure curves such as those of Figure 1 in Chapter I, one intuitively feels that alloys will fractionate when heated to evaporation temperatures, and this is indeed the case. The components of higher vapor pressure evaporate at a faster rate than the lower vapor pressure components resulting in a deposit of different composition than the original alloy. There are some measures that can be taken to counteract this problem to some extent but never fully. In a practical sense, this problem must be accepted as insurmountable. A large number of useful coatings can be obtained in which the composition of the deposit is somewhat different from the original evaporant alloy composition.

A somewhat similar situation exists with respect to compounds. Most compounds will partially dissociate upon evaporation, but will nevertheless often result in useful coatings. The most commonly used compounds in vacuum evaporation are two exceptional ones, MgF_2 and SiO), which do not dissociate upon evaporation.

Film Thickness Measurement and Control

Specifications for thin films to be vacuum deposited generally leave open the method of determining or measuring film thickness. The most basic film thickness measurement concept is depicted in Figure 7. We deposit a test film onto a test substrate that has been weighed prior to the deposition. We weigh the test substrate again after the deposition. The difference in weight is clearly the weight m of the deposited film. We know the length l and width w of the test substrate and hence of the deposited film. We know the type of material deposited so we can find the density d. The density multiplied by the volume is the weight of the film. The volume is the product of length l, width w, and thickness t. We thus have $m = lwtd$. All terms except thickness are known quantities so we can solve for t, the thickness of the film.

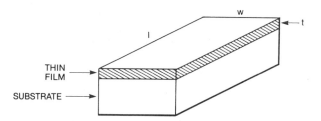

FIGURE 7   Thin film thickness measurement.

This method of determining coating thickness, referred to as the gravimetric method, is easily understood and is very simple in practice. It is unfortunate that this excellent method has come into some disrepute. In practice, results of the gravimetric method are almost invariably more accurate and more reliable than results obtained with the more esoteric equipment often preferred by many people. The disadvantages of this method are that the test substrate crowds out one or more payload substrates and that the results are determined only after the coating is completed. The gravimetric method is useful as a means of verifying that the correct thickness was deposited, but is not a means of process control.

Another means of measuring film thickness is by means of a stylus drawn across a step in a deposited film. The step can be generated either by masking a portion of a substrate during deposition or by etching a portion of the film after deposition. The displacement of the stylus as it is drawn over the step is detected by a transducer and recorded on a strip chart. The amplification and calibration problems are obvious. The problem of scratching or deforming the film surface detracts from the reliability of the measurement. The problem of reading the chart can be appreciated only by one who has gone through the process. The prime attraction of the stylus instruments is that the stylus measurement can be repeated frequently on a given film and can be repeated by other operators using other stylus instruments. The gravimetric method is more reliable and more accurate, but can be used only once on a given film. The real value of a stylus instrument is in measuring the thickness of a film for which the density is unknown.

Neither the gravimetric method nor the stylus method can serve as a process control monitor. These methods measure the film thickness after it has been deposited. In some cases, instead of using a control monitor, the film thickness can be controlled by totally evaporating a known charge of evaporant. Several trial runs are needed to establish the exact charge weight required to give the desired film thickness, but, once this is done, this provides repeatable coatings. This process control technique works well with MgF_2 and SiO for which other simple control methods do not work.

The thickness of metallic films can be easily monitored in process by monitoring resistivity. Typical elements for resistivity monitoring are shown in Figure 8. The glass substrate is coated through the mask to give a long, narrow conductive path between the conductive pads. A 0.2-cm-wide path with total length of about 20 cm provides for monitoring almost every thickness encountered for almost every metallic coating. The resistance of such a path is $R = 20\rho/0.2t$, or $R = 100\rho/t$, where t is the film thickness in centimeters and ρ is the resistivity of the material in ohm centimeters (Ω cm). In Table II we see that resistivities are in the range $(1 \text{ to } 100) \times 10^{-6}$ Ω cm so probe resistances will be in the range $R = (10^{-4} \text{ to } 10^{-2})/t$. Film thickness will be in the range 10^{-5}–10^{-4} cm resulting in probe resistances in the range 1–1000 Ω. Thin film resistivities are a

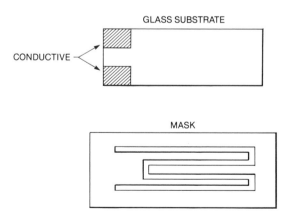

FIGURE 8   Resistance monitor.

TABLE II

Element	$\gamma(\times\ 10^5)(°C^{-1})^a$	$\rho(\times\ 10^6)(\Omega\ cm)^b$
Al	2.5	2.4
Cr	0.7	12.9
Cu	1.6	1.67
Au	1.4	2.35
Ge	0.5	46×10^6
Fe	1.2	9.7
Mg	2.5	4.5
Mo	0.5	5.2
Ni	1.3	6.8
Nb	0.7	12.5
Pd	1.2	10.8
Pt	0.9	10.6
Si	0.4	10
Ag	1.9	1.59
Ta	0.6	12.5
Ti	0.8	42.0
W	0.4	4.7
V	0.7	25
Zr	0.5	40
SiO_2	0.4	—
Stainless steel	1.7	76

[a] γ is the coefficient of linear thermal expansion (centimeters per centimeter per degree Celsius).

[b] ρ is resistivity. Values listed are for bulk material, not for thin films.

function of film thickness rather than fixed values as listed in Table II. As a consequence, the resistance monitor must be calibrated against the gravimetric method. Once this monitor has been calibrated, it provides an excellent process control monitor. Readout should be with a digital ohmmeter and should not be started until a visible coating has been deposited. It is best to have a switch to disconnect the ohmmeter from the monitor until a visible coating has been deposited.

Probably the most popular film thickness process monitor is the quartz crystal oscillator based on the piezoelectric properties of quartz. A thin quartz crystal wafer with conductive electrodes on its two surfaces is incorporated into an oscillator circuit just as in a fixed-frequency radio-frequency generator. This wafer is installed in a vacuum coating system so that one surface is exposed to the evaporation source. The coating deposited on the crystal surface loads the crystal and changes the oscillating frequency. The change in oscillating frequency is directly proportional to the added mass load and can be read directly in digital form with the proper electronic circuitry. Incorporating the evaporant density and proportionality factors into the electronic circuitry allows a direct readout of film thickness. Being able to read a thickness on a digital readout lends a great deal of credibility to the result, but much of this is somewhat questionable if not unwarranted. There are implicit assumptions that

(1) nothing has been done to violate the theoretical justification that thickness is directly proportional to frequency change,
(2) the quartz crystal has been cut properly,
(3) the electronic circuitry is calibrated properly, and
(4) the sensor is being used properly.

The quartz crystal thickness monitor is very useful, but it should be remembered that the basic calibration of this monitor is traced back to the simple gravimetric method and therefore should not be considered to be more valid than the gravimetric method. The quartz crystal monitor is an excellent process monitor for all materials, but its calibration, being based on the gravimetric method, should not be considered to give more reliable results.

Substrate Cleaning

The effectiveness of cleaning of substrates has a strong effect on the adhesion properties of deposited films. Actual measurement of the adhesion of a film is generally not possible. The universal qualitative measure of adhesion is the tape test using 3M Scotch brand #810 tape as specified in federal specification L-T-90. A piece of this tape is applied to a deposited film and pressed on firmly to ensure good contact and removal of trapped air. The tape is then pulled off with a firm, quick pull for the more severe test or with a firm, slow, even pull for the less severe test. A coating that passes the tape test will meet subsequent adhesion requirements, so this test, while only qualitative in nature, seems to differentiate between two very real classes of film-to-substrate bonding. Substrate cleanliness is the most important factor in achieving good adhesion.

The greatest threat to substrate cleanliness is in handling. One can convince oneself of this by washing and rinsing a microscope slide (as one might wash and rinse a household dish) and then holding the wet slide up out of the water for inspection. To avoid the great potential for suffering cuts, the eight edges and four corners of the microscope slide should be rounded off by use of a simple hone (obtainable from a hardware store). If one is wearing rubber gloves for this test the slide appears to remain clean, but if one is not wearing gloves, contamination is clearly visible, spreading from the fingers and covering the slide in a matter of seconds. We have been discussing atoms and molecules enough that by this time we can appreciate the problems caused by this contamination. This contamination layer is at least one atomic layer thick so evaporated atoms deposited onto such a contaminated substrate are actually separated from the substrate by the layer of contamination. There is no possibility that a coating can adhere to the substrate under such conditions. The fact is that much lower contamination levels interfere with film adhesion. Even the use of rubber gloves does not ensure against contamination, and the final stages of cleaning are best done using ceramic, glass, or metallic fixturing to hold parts and substrates.

In discussing cleaning of parts and substrates for vacuum coating operations, it is always assumed that the starting point is that at which there is no visible contamination. This should always be verified by wiping with a clean, white wipe, such as a Kimwipe. Assuming that the part passes this inspection, the first cleaning step for most parts and substrates is a vigorous scrubbing with detergent followed by vigorous scrubbing during the rinsing procedure. We use 18-MΩ deionized water for rinsing. After the scrub rinse, the part or substrate should be installed in a fixture for subsequent cleaning steps. The scrub rinse is followed by simple spray rinsing with deionized water. An isopropyl alcohol rinse removes the water to prepare the part or substrate for vapor degreasing. Most parts and substrates are ready for vacuum coating after vapor degreasing.

In some cases it may be found necessary to use ultrasonic cleaning or acid cleaning as additional steps. In these cases it is necessary to repeat the deionized water spray rinse followed by an isopropyl alcohol rinse and degreasing. In many cases, the cleaning process must then be followed by a baking process, which may be in air or in vacuum at temperatures ranging from 100°C to 1000°C depending on the substrate. Beyond this there are many cases where subsequent cleaning by ion bombardment is necessary.

It is hoped that this cursory treatment of substrate cleaning does not leave the impression that cleaning is a trivial matter. Cleaning is undoubtedly the most time consuming and most important part of all vacuum technology. In addition to cleaning parts, substrates, and fixturing, one must clean tools, handles, surfaces, and external fixturing. There is always a very high probability that someone will contaminate something that will subsequently contaminate something else that will ultimately contaminate a part or substrate.

There are many cleaning techniques and every type of part or substrate may require a different technique. The best technique to start with is the simple soap-and-water scrub, rinse, and degrease described here. Beyond that it is a matter of judgment and trial and error to find improved techniques for particular cases. There are some suggestions in the literature, but one's own ingenuity is usually the best guide as long as the simple scrub, rinse, and degrease method is not neglected.

As in previous chapters, there are innumerable facets that have not even been touched on in this discussion. Some of the most important types of evaporation sources were not even mentioned. Actual E-beam sources are much more complicated than the simple version described here. Film thickness measuring and monitoring systems using optical techniques were not mentioned here. Each section of this chapter could be greatly expanded. Nevertheless, it is felt that this chapter has provided a treatment of the basics that will provide the novice with a good grasp of the subject for his initial work and prepare him for further studies.

CHAPTER IV

SPUTTERING

If a solid or liquid is heated to a high enough temperature, it is possible for individual atoms to acquire enough energy via thermal agitation to escape from the surface. This means of causing ejection of atoms from a surface, called evaporation, has been our principal subject of discussion up to this point. We now want to discuss the process known as sputtering. If a solid or liquid at any temperature is subjected to bombardment by suitably high energy atomic particles (usually ions), it is possible for individual atoms to acquire enough energy via collision processes to escape from the surface. This means of causing ejection of atoms from a surface is called sputtering. Just as atoms ejected from a surface by evaporation can be used in depositing a coating on a substrate, atoms ejected from a surface by sputtering can be used in depositing a coating on a substrate. Any suitably energetic atomic particle impinging against a surface can cause sputtering. It is most convenient to accelerate ions to energies suitable for sputtering so we henceforth speak only of sputtering under ion bombardment.

Ion Bombardment of a Surface

The exact mechanisms by which atoms are ejected from a surface under ion bombardment are not known, but we can deduce some of the details of the interactions involved. An ion is essentially the same size as an atom, so when an ion impinges against a surface, it actually collides initially with a surface atom. At energies at which significant sputtering occurs, the energy exchange between an incident ion and a surface atom is much greater than either lattice binding energies or vibrational energies of lattice atoms. As a consequence, neighboring atoms do not become involved in the primary collision, so the primary collision is strictly binary, with the incident particle giving up a significant fraction of its primary energy to the struck atom and retaining the remaining fraction.

It is most common in sputtering that the ions are incident to the bombarded surface in the direction parallel to the surface normal. In this case, if the mass of the incident ion is lower than the mass of the surface atom with which it collides and if the collision is head-on or nearly so, the incident ion will bounce back away from the surface. This is the case depicted in event A of Figure 1. The surface atom involved in such a collision will be driven in a direction toward the surface interior. All other cases are similar to event C of Figure 1, where both the incident ion and the struck atom leave the

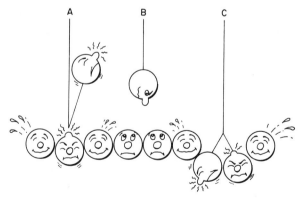

FIGURE 1   Ions at normal incidence onto a surface.

point of collision along paths directed toward the interior of the surface. If the incident ion is of greater mass than the struck atom, both ion and atom will leave the point of collision in directions toward the surface interior regardless of whether the collision is head-on or glancing. We thus have at least one and usually two particles traveling toward the interior of the surface at energies that are less than the primary energy of the incident ion but still much greater than lattice energies. As can be seen from Figure 2, no atom will be ejected from the surface as a direct result of the primary collision. For a surface atom to be ejected, it must acquire a velocity component in the direction opposite to the direction of the original velocity of the incident ion. As indicated in event C Figure 2, the greatest possible angle between the original momentum vector of the ion and the subsequent momentum vector of the struck atom is 90°, and in this case the momentum vector or the velocity of the struck atom is zero. The surface atom therefore cannot acquire a velocity component in the direction away from the surface as a direct result of the primary collision. The direct result of the primary collision will be at least one and usually two secondary binary collisions inside but still close to the surface.

G. K. Wehner, who in the 1950s established the framework that put sputtering on a solid scientific basis, likes to speak of sputtering as a game of three-dimensional billiards played with atoms. Using such an analogy, we can see that it would be possible for atoms to be ejected from the surface as a direct result of the second set of binary collisions. Considering event B of Figure 2 we can see that it should be possible for either the ion or the struck atom to leave the point of impact at an angle (with respect to the original ion velocity direction) greater than 45°. It should then be possible for a secondary collision in the same plane of motion to result in a lattice atom leaving the point of secondary impact at an angle greater than 45°. Two angles greater than 45° add up to an angle greater than 90° so this lattice atom has a velocity component directed outward from the surface and thus has the potential of being ejected. Reflecting further on this, we can see that such atoms could not be ejected parallel to the surface normal; i.e., such atoms could not be ejected in the direction opposite to the direction of the incident ion. This would require two 90° deflections, at least one of

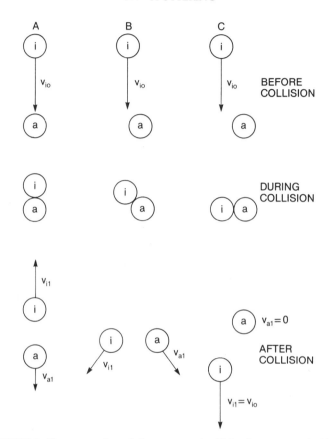

FIGURE 2 Representation of three types of collision between an incident ion and a surface atom. The mass of the incident ion is represented here as being less than the mass of the struck atom.

which would involve a lattice atom being deflected at 90° and acquiring zero velocity in the process. An atom with zero velocity can neither cause ejection nor be ejected. If we carry this analogy too far, we might conclude that sputtered atoms might be ejected most strongly in directions away from the surface normal, but this is not the case. It has been found that, in the case of normal ion incidence at energies of interest to us, sputtered atoms are ejected from the surface in essentially a cosine distribution, the same as evaporated atoms. This is interesting in that the most probable direction of ejection is in exactly the opposite direction to the direction of the incident ion. Evidently the energy brought in by the incident ion is so randomly distributed by multiple collisions before ejection of an atom that the incident momentum vector is completely lost and plays no part in the ejection. We must keep in mind that this result is only for the case of ion bombardment at normal incidence.

In some situations, bombarding ions may be incident on the surface at oblique angles. In this case, there is a high probability for sputtering to result from the primary collision between the incident ion and the surface atom with which it first collides. In the case of oblique incidence, the incident momentum vector is found to play a strong role in the ejection pattern, and sputtered atoms are ejected very strongly in the forward direction. The sputtering yield, i.e., the number of atoms ejected per incident ion, may be as much as an order of magnitude greater in the case of oblique incidence than in the case of normal incidence of the bombarding ion. It is probable that this results partly from partial retention of the original momentum vector in subsequent collisions and partly from the fact that these subsequent collisions occur closer to the surface than in the case of normal ion incidence.

Many phenomena can occur as a result of ion bombardment of a surface, depending on the surface, the ion, the energy, and many other factors. We have touched on a few details of one of these phenomena, i.e., sputtering, as a means of introducing this subject. There has been quite a bit of good theoretical work in this field, but no one has yet developed a comprehensive theory of sputtering, and no one seems likely to in the near future. Lacking a theoretical structure for discussion it will be best for us to consider the experimental aspects of sputtering. We first want to touch on another of the phenomena that occur as a result of ion bombardment of a surface, namely, secondary electron emission.

When an ion approaches a surface at any energy and at any angle, it will be neutralized before impact by interaction with the lattice electrons of the surface. This interaction occurs within less than an atomic diameter of the surface and actually involves two lattice electrons. One lattice electron is captured by the ion as an orbital electron, thus neutralizing the ion. The second electron acquires the excess energy and momentum given up by the neutralizing electron and may, as a result, be ejected from the surface. The probability of ejection depends on the type of ion and the type of material making up the surface, but in all cases is quite low, usually of the order of a few percent, and is usually not known very accurately. The Greek letter γ (gamma) is used as a symbol to represent the probability of ejection. This ejection process is referred to as secondary emission, and the electrons ejected in this process are referred to as secondary electrons.

Glow Discharge

Sputtering was first reported in 1852 by W. R. Grove, who observed metallic deposits on glass walls during studies of electrical conductivity of gases. The origins of many branches of science can be traced to phenomena first observed in studies of electrical conductivity of gases. Among these are cathode rays (electrons), positive rays (ions), x rays, cosmic rays, and, of course, sputtering. Here we consider the aspects of electrical conductivity of gases that relate to sputtering.

There have been many very ingenious methods employed in studies of the electrical conductivity of gases, but the simple arrangement shown in Figure 3 serves best to show how these studies can lead to the observation of sputtering. This arrangement

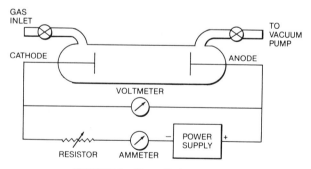

FIGURE 3 Glow discharge tube.

provides for several variable parameters: type of gas, pressure, voltage, current, electrode material, electrode dimensions, electrode spacing, and ratio of electrode dimensions to electrode spacing. We shall consider only a limited number and range of these parameters. The gas will be argon in the pressure range below 1000 mTorr. The electrodes will be disks of 5-cm diameter and 15-cm spacing.

We first want to determine how far an electron can travel in this simple system. We want to calculate the mean free path of an electron in a gas in exactly the same way that we calculated the mean free path of a gas atom in gas. We recall that we earlier calculated the diameter of an argon atom and found a value close to the handbook value of 3.82×10^{-8} cm. Using the handbook value and Loschmidt's number, 2.69×10^{19} gas atoms/cm^3 at normal temperature and pressure (760 Torr), we calculate that at a pressure of 1000 mTorr there are 3.54×10^{16} gas atoms/cm^3, so each argon atom on the average is within a volume of 2.82×10^{-17} cm^3, which we think of as a cube of dimension 3.04×10^{-6} cm. We think of the gas (argon) as being in ficticious layers 3.04×10^{-6} cm thick. We think of each layer as being divided into (3.04×10^{-6})-cm squares with each square partially blocked by an argon atom of diameter 3.82×10^{-8} cm or of area 1.15×10^{-15} cm^2. The area of the square blocked by each argon atom is 9.29×10^{-12} cm^2. The ratio of the area of the square to the area of the argon atom is 8000 to 1, which we recall as being the important factor in determining the mean free path. In order to calculate the mean free path of a gas atom we had corrected this number by including the size of the gas atom that we considered to be in motion. Here the moving particle is an electron, which we think of as being a dimensionless electrical charge, so the 8000 to 1 ratio needs no correction. The probability than an electron will strike a gas atom while passing through a given layer is thus 1 chance in 8000. The average electron will have made a collision with an argon atom by the time it has passed through 4000 layers so the mean free path of an electron in argon at 1000-mTorr pressure is $4000 \times 3.04 \times 10^{-6}$ cm, or 1.22×10^{-2} cm. We recall that this includes even the most trivial collisions and that we had coined the term mean deflection path, ten times the mean free path, as being the path length for a meaningful collision. The

mean deflection path for an electron in argon at 1000 mTorr is thus
0.122 cm. If this is the mean deflection path, there must be a
significant number of paths of length in the range 0.5–1 cm. A
more sophisticated derivation gives values about twice as great as
ours, but this merely adds some assurance that there will be a
significant number of electrons that will traverse paths in the 1-cm
length range before being slowed down by a collision. In addition
to this we are ignoring some other effects that may change our
results somewhat but cannot negate them.

We want here to calculate the mean free path of a singly charged
argon ion in argon gas at 1000 mTorr. We can find a handbook
value of 1.54×10^{-8} cm as the radius of an argon ion, and we have
3.82×10^{-8} cm as the diameter of an argon atom. An argon atom
can therefore block an argon ion out of an area of $\pi(1.54 \times 10^{-8}$ cm
$+ 1.91 \times 10^{-8}$ cm$)^2$, or 3.74×10^{-15} cm^2. The ratio of the area of
the square containing an argon atom to the blocked area is thus
2470 to 1, so the mean free path of an argon ion is estimated to be ½
$\times 2470 \times 3.04 \times 10^{-6}$ cm, or 3.75×10^{-3} cm, under these conditions. The mean deflection path is 3.75×10^{-2} cm. There are not as
many argon ions having paths in the 1-cm range, but there must
still be some.

We now consider the system of Figure 3 with argon gas at 1000
mTorr and a dc voltage of 1500 V. There are no charge carriers
(electron or ions) in the argon gas so there is no current flowing,
and the full 1500 V is dropped uniformly across the 15-cm space
between the electrodes. If an electron is introduced into this space,
it will be accelerated away from the cathode and toward the anode.
There is a high probability that before this electron reaches the
anode it will have had at least one path segment in which it will
have traveled a distance of about 1 cm without making a significant
collision with an argon atom. With an electric field of 1500 V per 15
cm (i.e., 100 V/cm), this electron will have acquired 100 V of energy
while traversing the 1-cm path segment. There is a high probability
that this (now energetic) electron will subsequently make a significant collision with an argon atom. In such a collision, the free
electron will give up energy to an orbital electron of the atom. If
this energy is less than the ionization potential, the orbital electron
will be excited to a higher energy level for about 10^{-8} sec and will

then return to the ground state with the emission of one or more photons of light. If this energy is greater than the ionization potential, the orbital electron will separate from the atom becoming a free electron and leaving a free ion. The second free electron encounters the same electric field and reacts in the same manner as the first free electron. The first free electron, after the collision, continues to react to the electric field and will continue to acquire energy from the field and give it up to orbital electrons of atoms with which it collides. There is here the potential for a cascading of the generation of free electrons and free ions.

We can see that if only the electrons were involved in generating new charge carriers, the current would build up to a maximum and quickly decay to zero. This is because new charge carriers would be generated closer and closer to the anode. When the last free electron is collected by the anode, there is no further mechanism for generating new charge carriers. The ions are able to provide such a mechanism but not directly. Ions must have much higher energies than electrons to ionize gas atoms, but are unable to acquire as much energy due to having much shorter mean free paths. The ions, being positive in charge, are accelerated toward the cathode, but, due to frequent collisions with gas atoms, seldom acquire energies as high as energies acquired by electrons. Even the occasional ion that acquires as high an energy as electrons acquire cannot generate charge carriers directly because an ion–atom collision at these energies does not cause ionization. As mentioned previously, when an ion of even this low energy approaches very near the cathode surface, within a distance of less than an atomic diameter it becomes neutralized by interaction with the lattice electrons of the cathode. Conservation of momentum and energy considerations are satisfied in this interaction only if a second lattice electron besides the neutralizing electron is involved in this interaction. This second electron acquires energy and momentum in the interaction and has a probability of a few percent (depending on the type of ion and the type of material making up the cathode surface) of being ejected. An electron ejected from the cathode surface in such a manner is referred to as a secondary electron, and the Greek letter γ is used as a symbol to represent the probability of ejection. Such an electron, once ejected from the cathode surface, encounters the electric field between the cathode and the anode and reacts in the same manner as the other free electrons described previously. Such an electron is therefore accelerated toward the anode, acquiring energy and making collisions along the way, thus adding to the number of free electrons and free ions in the region between the anode and the cathode. If an electron can generate enough ions as it traverses the tube that the number of ions generated multiplied by γ is equal to or greater than 1, then it will be replaced by one or more secondary electrons, and the discharge will be self-sustaining. This condition will be met in the system under discussion once the discharge is initiated, and this actually occurs as soon as the voltage is applied. The first free-electron–free-ion pair is probably generated by a cosmic ray passing through the tube. The electron velocities in this discharge are so high and the interactions occur so quickly after the formation of the first free-electron–free-ion pair that the generation of charge carriers (free electrons and free ions) cascades up to an equilibrium condition instantaneously. This equilibrium discharge condition is known as a glow discharge. At equilibrium, the voltage drop between the anode and the cathode is about 150 V, and the discharge current builds up to the point that the voltage drop across the current limiting resistor is equal to the difference between this and the supply voltage. If the supply voltage is increased or the current limiting resistance is decreased, the current increases, but the anode to cathode voltage does not change significantly.

It is not necessary to measure current and voltage to know that there is a glow discharge because the discharge does indeed glow. So many of the electron–atom collisions are of the type that result in excitation of an orbital electron to a higher state with subsequent decay to the ground state and emission of photons that there is a great deal of light emitted by the glow discharge. The general appearance of the discharge is depicted in Figure 4 and is pretty much the same for a wide range of gases, pressures, and currents. There is a glowing column that nearly fills the tube, extending from the anode nearly to the cathode. This column (known as the positive column) ends in a convex shape with the adjacent space appearing dark. Closer inspection reveals that this space (known as the Faraday dark space) is luminous but much less so than the

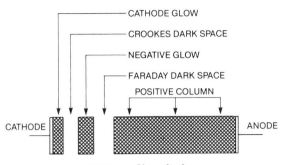

FIGURE 4 Glow discharge.

positive column. Further away from the end of the positive column, the luminosity increases again, becoming more intense than the positive column. This region (known as the negative glow) is again defined on the cathode side by a space (known as the cathode dark space or the Crookes dark space) that is of much lower luminosity. Finally there is a thin sheet of higher luminosity (known as the cathode glow) that appears to cling to the cathode surface. There are qualitative explanations for some of these regions but not for all. These regions are usually not clearly defined even on the cathode side and never on the anode side.

Before the initiation of a discharge, the electric field between the anode and cathode is uniform; i.e., the voltage drop between the anode and cathode is distributed uniformly. This is not the case after the discharge starts. It is found that the electric field in the glow discharge is uniform but quite low over the length of the positive column and into the Faraday dark space. The electric field from there into the negative glow decreases nonuniformly, even taking on small negative values. The highest and most nonuniform electric field is found in the region starting at the brightest part of the negative glow and running to the cathode. The greatest part of the voltage between the anode and cathode is dropped across this region, and therefore the charged particles (electrons and ions) experience their greatest acceleration in this region.

We can see that this can be associated with the decreased luminosity in the Crookes dark space. Because of the high electric field, electrons travel through this region at high speeds. If a collision occurs to take energy away from an electron and slow it down, the field quickly accelerates the electron back to high speed so no electron remains at low speed in this region and no electron spends much time in this region. In fact, it is found that the electron density in the Crookes dark space is lower than elsewhere in the discharge. Since there are not as many electrons here, there are not as many electron–atom collisions here. Since there are not as many collisions here, there are not as many orbital electrons excited to higher energy states from which they decay to the ground states by photon emission. Thus there are not as many photons emitted here, or, in our original wording, the luminosity is lower here. Another factor involved is the fact that the electrons here have higher energy, and high energy electrons have a lower probability of taking part in this type of excitation collision. Lower energy electrons, which have a higher probability of causing this type of excitation, do not remain at low energy because of the high electric field that exists here.

We can speculate on two possible causes of the cathode glow. One possibility is that the electrons ejected from the cathode surface as a result of the interaction between incident ions and lattice electrons, being at low energies near the cathode surface, are experiencing excitation collisions with gas atoms near the cathode surface. Another possibility is that neutralized ions may initially acquire the orbital electron in an excited state and emit photons to decay to the ground state.

It is interesting that the distance from the cathode to the negative glow region is approximately ten electron mean free path lengths, the distance for which we coined the term mean deflection path. Most electrons from the cathode have experienced a significant collision with an atom by the time they reach the negative glow region (or at least before they have passed through it) and have given up a great deal of energy. The electric field is lower here so they are not greatly accelerated and therefore do not acquire a great deal of additional energy. Electrons therefore spend more time in this region and therefore experience more collisions with atoms in this region. Since these electrons have lower energies, these collisions are more likely to be of the type which leads to excitation and subsequent emission of photons. We observe this as greater luminescence.

The electron–atom collisions that result in ionization occur predominantly in the cathode dark space and the negative glow region, where the elctron energies are highest. The electric field in most of this region is such that the ions are accelerated toward the cathode and the electrons toward the anode. The rate at which ions reach the cathode results in a current i^+. As these ions are neutralized, a fraction γ are involved in interactions that result in the ejection of secondary electrons from the cathode surface. The rate at which electrons reach the anode must result in a current i^-, equal to the sum of the ion current i^+ plus the current γi^+ resulting from the secondary electrons. The ammeter, of course, measures i^-, which is equal to $i^+ + \gamma i^+$ or $i^+(1 + \gamma)$. The factor $(1 + \gamma)$ is one we shall encounter frequently. The ions involved in i^+ are generated in the cathode dark space and in the negative glow region, as are most of the electrons involved in i^-. The secondary electrons $(i^- - i^+)$ are generated at the cathode surface.

The Faraday dark space is less luminous because, even though the electron density is higher, the electrons here have too little energy to cause excitation of the atoms. Most electrons reaching this region have already lost most of their acquired energy in collisions with atoms and, in addition, have passed through a small retarding field, which further reduced their energy.

In the positive column there is a small electric field that can accelerate electrons up to energies sufficient to cause excitation and even ionization, so the luminosity is higher. The average ion density in this region is equal to the average electron density, so the average charge density is zero. This condition is referred to as a plasma. Technically, any gas could thus qualify as a plasma, but we usually feel that the ion and electron densities should not be zero in what we call a plasma.

As was indicated earlier, the ions impinging against the cathode in this discharge are not energetic enough to cause sputtering. This is because the mean deflection path (our designation and a term not used elsewhere) of an ion at this gas pressure is so short that collisions prevent ions from accumulating any appreciable energy in the electric field. We know that we can increase the mean deflection path by decreasing the gas density, which in turn we know we can do by decreasing the gas pressure. As the pressure is decreased, the positive column decreases in length, and both the negative glow and the Faraday dark space increase in length. The current decreases, and the voltage drop from anode to cathode increases as pressure decreases. At pressures of approximately 100 mTorr sputtering becomes detectable. The electric field is high enough and the mean deflection path of an ion long enough for some ions to accumulate enough energy to cause sputtering when they impact on the cathode surface. This is not an efficient arrangement for sputtering because the mean deflection path is still so short that it strongly restricts the ions in accumulating energy and strongly restricts the movement of sputtered atoms away from the cathode. Nevertheless, this type of system, with some refinements, was used for sputtering studies until 1930 when A. Guentherschulze and K. Meyer introduced the triode sputtering system. We shall see later how this system operates.

Our reaction to the problem of ion mean deflection paths still being too short is to pump to even lower pressure, but we find that at lower pressures the discharge is extinguished. We recall that the requisite for a self-sustained glow discharge is that an electron ejected at the cathode must generate enough ions before reaching the anode to ensure ejection of another electron at the cathode. This requires that the electron mean deflection path be long enough to allow the electron to accumulate ionizing energy but short enough to assure enough electron–atom collisions to generate the needed number of ions. At pressures much below 100 mTorr, the electron mean deflection path is so long that electrons do not make enough ionizing collisions to sustain a glow discharge.

There is little to be gained by further consideration of sputtering in a glow discharge. Our discussion here has omitted many details and many variations. The glow discharge is not a simple phenomenon. The understanding of the glow discharge is far from complete, and the explanation of even the portions of it that we have undertaken here is inadequate. We have attempted only to cover enough to bring out some of the concepts that are important in the pursuit of experimental studies of sputtering.

The glow discharge extinguishes at pressures below 100 mTorr because there are not enough secondary electrons released at the cathode to sustain the discharge. The discharge can be sustained if another source of electrons besides secondary emission is used. If the cathode of Figure 3 is replaced by a wire heated to a temperature high enough to emit electrons, the discharge can be maintained at much lower gas pressures. This discharge is spoken of as a supported discharge. The plasma generated in this discharge has the characteristics of the positive column of the glow discharge. The voltage drop is low, so the electric field is low. The electrons are accelerated to energies high enough to cause excitation and ionization, so the discharge is luminous just as the positive column of a glow discharge is luminous. The ions are accelerated to energies very near the full voltage drop between anode and cathode, but these energies are much too low to result in sputtering. The difference between this case and the case of the higher pressure glow discharge is in the factor that limits the ion energy. In this case, the available voltage is too low for sputtering. In the case of the higher pressure glow discharge, the available voltage is high enough, but the frequent collisions with gas atoms take energy away from the ions before they are able to acquire enough energy to cause sputtering.

The Langmuir Probe

Glow discharges had been studied by I. Langmuir, using a small metal probe inserted in the discharge path. The same probe can be used to study a supported discharge. When a negative potential (voltage) is applied to such a Langmuir probe, an electric field is established in the vicinity of the probe such that ions within the boundary of the field are accelerated to the probe but electrons are repelled and are unable to enter the field. Because of the lack of electrons in this region there is no excitation and no photon emission so the region is clearly visible as a dark region reminiscent of the cathode dark space in a glow discharge. This region is referred to as an ion sheath. In essence, the positive ion content of this region is just that needed to counterbalance the negative potential on the probe so that no field due to the probe is detectable in the plasma beyond the edge of the ion sheath. The plasma is thus not disturbed or changed by the probe when a negative voltage is applied to the probe. The thickness of the ion sheath is a function of the ion density of the plasma and of the voltage applied to the probe. The dependence on either factor is not very strong, so wide variations in applied voltage do not change either the ion sheath thickness or the ion current a great deal. Every electron reaching the edge of the ion sheath is reflected, and every ion reaching the edge of the sheath is accelerated to the probe. The number of ions reaching the edge of the ion sheath per square centimeter each second is given by the now very familiar expression $\frac{1}{4}n^+v^+$, where n^+ is the ion density (i.e., the number of ions per cubic centimeter), and v^+ is the average ion velocity.

Although we shall find that our main interest in the Langmuir probe is in the case of a negative potential, it is of value and interest to consider the consequence of reducing the negative potential through zero and applying a positive potential. A plot of probe current versus probe voltage obtained in this way is depicted in Figure 5. When we speak here of the potential applied to the probe, we mean the potential with respect to the plasma, but in a practical sense we must apply the potential from the power supply between the probe and another electrode. We could choose the anode, the cathode, or ground as the reference electrode. Experience has taught us that the plasma assumes a potential near anode

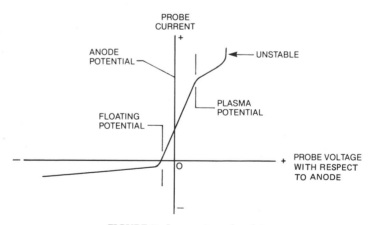

FIGURE 5 Langmuir probe plot.

potential so we always connect the power supply between the probe and the anode. The plot depicted in Figure 5 is drawn on this basis; however, in our discussion we shall continue to refer to the probe potential as being with respect to the plasma.

A negative potential applied to the probe results in an ion current to the probe and the formation of an ion sheath around the probe. The current to the probe is negative just as it would be in a common electrical circuit, but, unlike a common electrical circuit, the current does not change a great deal as the applied voltage is changed. The probe current depends on the plasma density, which is independent of the probe voltage. Random motion of ions within the plasma (exterior to the ion sheath) results in ions entering the boundary of the sheath at a rate ($\frac{1}{4}n^+v^+$ ions/cm^2)/sec. This results in a probe current (negative) of $\frac{1}{4}n^+v^+A^+$, where A^+ is the surface area of the ion sheath. The diameter of the ion sheath (and thus its area) is weakly dependent on the applied voltage so the probe current assumes somewhat greater negative values at more negative probe voltages. The slope of the probe curve in the negative voltage range is therefore slightly positive.

As the negative probe potential is reduced to a value near zero, it reaches a value at which the higher energy plasma electrons are able to penetrate the ion sheath and reach the probe. When the voltage is such that the current due to these electrons is equal to the ion current, the net current, which is the current we read on a meter, is zero. Under this condition, the probe potential is exactly the same as the potential that an electrically isolated probe or an insulated surface would assume. This voltage is referred to as the floating potential.

As the probe voltage approaches closer to zero from this point, the probe current becomes dominated by the electron contribution, which rises rapidly in this range. In this region we have the anomoly that the applied voltage is negative but the probe current is positive. When the probe reaches plasma potential, the ion sheath has disappeared, and the plasma extends all the way to the probe surface. The probe receives random ion current at a rate determined by $\frac{1}{4}n^+v^+$ and random electron current at a rate determined by $\frac{1}{4}n^-v^-$. We know that n^+ and n^- are equal in a plasma. Even if the electrons and the ions were in temperature equilibrium with each other, v^- would be much greater than v^+ In either a glow discharge or a supported discharge, the electron temperature is higher than the ion temperature so this difference is magnified, and we find that the electron current is about two orders of magnitude greater than the ion current. Thus, when the probe is at plasma potential, the net current received by the probe is dominated by the electron contribution.

When the probe potential becomes positive with respect to the plasma, electrons are accelerated to the probe, and an electron space-charge sheath (from which ions are now repelled) forms around the probe. The change in electron sheath thickness with probe voltage is noticeable as an increasing current with increasing probe voltage. When the probe voltage reaches the ionization potential of the gas atoms, the probe current increases rapidly but in a very unstable fashion that cannot be depicted in Figure 5 except by an upward turn of the plot.

The Triode Sputtering System

We are now in position to consider the triode sputtering system introduced by Guentherschulze and Meyer and later modified by Wehner. A system of this type is depicted in Figure 6. The vacuum pumping system (not shown here) is able to maintain an ultimate

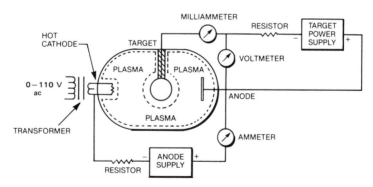

FIGURE 6 Triode sputtering system.

pressure at least in the 10^{-7} Torr range to minimize contaminant problems. After a good vacuum is obtained, the discharge gas (usually argon) is bled in at a rate at which the mechanical pump can maintain a foreline pressure that is not detrimental to the diffusion pump. A state of dynamic equilibrium is easily reached in which argon is being pumped out at the same rate at which it is being admitted to the chamber. It is common to have a pumping system in which, in this state of dynamic equilibrium, the chamber pressure is 1 mTorr and the foreline pressure about 50 mTorr. This is a very convenient and easily achieved operating pressure, but the early systems and many present-day commercial systems are designed to operate at chamber pressures of 5 mTorr or more. It is not possible to admit argon at a higher rate to increase the chamber pressure because this would be detrimental to the operation of the diffusion pump. It is therefore necessary to throttle down the high vacuum valve to raise the chamber pressure. This reduces the pumping speed and has the detrimental effect of allowing the contaminant level in the chamber to increase in proportion to the argon pressure. We prefer to operate at 1 mTorr so that the pumping need not be throttled down and the cleanest possible vacuum is maintained. In addition to resulting in a cleaner vacuum, operation at 1 mTorr is found to be more convenient. The pressure of 1 mTorr is easily set by monitoring with the Bayard–Alpert ion gauge, which is linear and very responsive at this pressure. At pressures of 5 mTorr and higher, the Bayard–Alpert gauge becomes nonlinear and not very precise in normal operation. If operated at an emission current of 50 μA, the Bayard–Alpert gauge is fairly reliable at mTorr, but this requires a modified ion gauge control circuit. There are special purpose gauges and gauge controls designed to operate at these higher pressures, but one must exercise a great deal of care in using and interpreting these gauges.

With the discharge gas adjusted to the desired operating pressure, the cathode is electrically heated to a temperature at which it emits electrons to support a discharge. When power is applied to the anode, electrons are accelerated toward the anode, argon atoms are excited or ionized by electron collisions, and the entire chamber is filled with plasma. The target, like a large Langmuir probe, may have negative dc voltage of any value applied to it without affecting the plasma other than within the ion sheath that forms around the target. Every ion reaching the edge of the sheath is accelerated to the target and has a probability of causing ejection of atoms from the target surface, i.e., sputtering. Every electron reaching the edge of the sheath is reflected with the consequence being that there is no excitation and no photon emission within the sheath so that the ion sheath is clearly visible as a dark region surrounding the target. The ion current to the target is a measure of the number of ions per second bombarding the target. Since the charge of an ion is 1.6×10^{-19} C (coulomb) and an ampere is 1 C/sec, the target current in amperes divided by 1.6×10^{-19} gives the total number of ions per second that have struck the target. We measure the target current and use it directly in calculating sputtering yields, but we are also interested in the target current density, i.e., the target current divided by the target area.

Current density in an argon plasma of the type usually encountered in sputtering ranges from a few tenths of a milliampere to a few milliamperes per square centimeter. It is interesting to calculate the maximum possible current density based on 100% ionization. We first calculate the rate at which argon atoms strike surface walls at operating pressure, which we take as 1 mTorr. We remember that we found that this rate is given by the expression $\frac{1}{4}nv$, where n is the numerical density of argon atoms and v is the average velocity of argon atoms. We remember Loschmidt's number, 2.69×10^{19} gas atoms/cm^3 at a pressure of 760 Torr. We calculate from this that the numerical density of argon atoms at 1 mTorr (10^{-3} Torr) is 3.54×10^{13} atoms/cm^3. We recall from Chapter III that the velocity of a gas atom is given by $v = 1.58 \times 10^4 \sqrt{T/M}$. Taking the temperature as 300°K and the atomic weight of argon as 39.9, we calculate that $v = 4.33 \times 10^4$ cm/sec. We can now calculate $\frac{1}{4}nv$ and find that interior chamber surfaces are struck by argon atoms at the rate 3.83×10^{17} atoms/cm^2/sec. If we have 100% ionization, then each impact results in a charge transfer of 1.6×10^{-19} C, the charge of a singly charged ion. The rate of charge transfer under 100% ionization is therefore $3.83 \times 10^{17} \times 1.6 \times 10^{-19}$, or 6.13×10^{-2} C/cm^{2}sec. A charge transfer of 1 C/sec is 1 A so the current density is 6.13×10^{-2} A/cm^2, or 61.3 mA/cm^2. We thus see that normal sputtering plasmas achieve an ionization of from a fraction of a percent to a few percent.

A question that sometimes arises is whether or not there is a cumulative effect involved in sputtering, i.e., whether the effects of an ion impact will dissipate before another ion strikes that disturbed area. This is a rather complicated question and even a first-approximation estimate involves ideas that are somewhat outside of the intended scope of this book. We have already noted that the primary impact of an ion against a surface atom must be considered as an independent binary collision. If we consider all subsequent secondary impact events to be also independent binary collisions, we can find that 20 collisions are sufficient to dissipate particle energies to well below ejection levels. An ion impact therefore affects an area of radius r equal to 20 atomic diameters; i.e., $r = 20 \times 2.5 \times 10^{-8}$ cm $= 5 \times 10^{-7}$ cm. This area is $A = \pi r^2 = 25\pi \times 10^{-14}$ cm^2. At the lowest velocity involved (5×10^4 cm/sec), the outer periphery of this area would be reached in $(5 \times 10^{-6}$ cm$)/(5 \times 10^4$ cm/sec$) = 10^{-10}$ sec, so a second ion impact must occur within 10^{-10} sec after the first one in order for a cumulative effect to occur. The second ion must strike the $(25\pi \times 10^{-14})$-cm^2 area affected by the first ion so the impact rate must be 1 ion/$25\pi \times 10^{-14}$ cm^2 each 10^{-10} sec, or 1.27×10^{22} impacts/cm^2 sec. We have just seen above that the highest impact rate possible at 100% ionization is 3.83×10^{17} impacts/cm^2 sec. We can therefore conclude that there is no cumulative effect involved in sputtering in the sense that excess energies imparted to surface atoms as a result of ion impact are dissipated before another ion impact occurs within the affected area.

The triode sputtering system depicted in Figure 6 meets the six criteria for conditions under which reliable sputtering results may be obtained. These criteria were, of course, derived from the works of many investigators, but they were first recognized as a complete entity and set down as such by Wehner.

1. Low enough gas pressure to avoid back diffusion of sputtered atoms.

2. Low enough gas pressure that ions will not collide with gas

atoms during their acceleration to the target. The ions incident on the target thus have a well-defined energy.

3. The ion accelerating voltage (and thus the kinetic energy of the ions) should be controllable independent of other parameters such as gas pressure and plasma density.

4. The excitation electron energy should be low enough that formation of multiply charged ions is negligible.

5. The target should be large compared to the thickness of the ion sheath. Target edges and corners should be cylindrical or spherrical sections with radii of the order of the thickness of the ion sheath.

6. Plasma density should be high relative to background gas impurities.

We have introduced a new term, the target, in the second and the fifth criteria, but it is unlikely that anybody has been very surprised by it. It is quite natural to speak of a bombarded object as a target, and it is likely that everybody has recognized that this object is going to be bombarded by ions and is therefore called a target.

We discussed the concepts of the first two criteria in the glow discharge section. The concept of the third criterion is implicit in the discussion of the Langmuir probe at negative potential. The fourth criterion is new to us. It happens that if an electron of energy sufficiently greater than the first ionization potential of an atom collides with an atom, then more than one electron can be separated, leaving an ion charged doubly or triply. A doubly charged ion can acquire twice as much energy in a given electric field as a singly charged ion. The presence of a significant number of doubly charged ions would result in a target being bombarded by ions of two different energies, and data taken under this condition would be meaningless. We take up the fifth and sixth criteria in later parts of this section.

The gas inlet and the pumping system have been omitted from Figure 6. The vacuum chamber and feedthroughs have been vaguely represented as an outline separating the vacuum region from the atmospheric pressure region. The cathode, represented as a resistance, is heated to electron emission temperature by a low voltage ac supply. The anode is a metal disk of about 10-cm diameter. The electron density is usually so high that the plasma assumes a potential a little bit more positive than the anode so that there is a slight retarding voltage applied to electrons reaching the anode. This is because the random electron impingement on the anode, given by $\frac{1}{4}n^-v^-$, would otherwise result in an electron current, given by $\frac{1}{4}e^-n^-v^-$, greater than the normal operating parameters require. We have here introduced the term e^-, equal to 1.6×10^{-19} C, the charge of an electron. We had earlier introduced e^+, equal to 1.6×10^{-19} C, the charge of a singly charged ion. These two charges, equal in magnitude and opposite in sign, are encountered often in the coming sections.

The target, the hot cathode, and the chamber are all at negative potential with respect to the plasma and therefore have ion sheaths surrounding them. The target, being at the greatest negative potential, has the thickest ion sheath; the chamber wall, being at the least negative potential, has the thinnest ion sheath. The entire anode-to-cathode voltage is dropped across the ion sheath around the hot cathode so all of the electrons acquire ionizing energy before reaching the plasma. Ionizing collisions therefore can occur throughout the plasma beginning at the edge of the ion sheath surrounding the cathode. In collisions with gas atoms, ionizing electrons give up part of their energy and are deflected in randomly different directions throughout the chamber. As a consequence, ionizing electrons reach all parts of the chamber to make ionizing collisions, and the entire chamber is fulled with plasma. The target, immersed in the plasma like a large Langmuir probe, does not affect the plasma except within the ion sheath around the target.

This system is usually operated at a pressure of about 5 mTorr, resulting in an ion mean free path length of about 1 cm, or an ion mean deflection path length of about 10 cm. The ion sheath thickness is only a fraction of a centimeter so the second criterion is met. The mean deflection path of a sputtered atom is about the same as that of an ion so the first criterion is met. The target does not alter the plasma so any negative voltage can be applied to the target; therefore the third criterion is met. The anode-to-cathode voltage is in the 25–35-V range so the fourth criterion is met. Target sizes are of the order of several centimeters, and targets are spherical or cylindrical with hemispherical ends so the fifth criterion is met. The sixth criterion is a vacuum-dependent factor, but, remembering Chapter III, we can be sure that our vacuum will be good enough that this criterion is met.

The triode sputtering system shown in Figure 6 is not highly effective for depositing coatings and was not really designed to deposit coatings. This system was designed to measure sputtering yields (the number of atoms ejected per incident ion) by the gravimetric method. The target is sputtered at a fixed ion energy for a long enough time to remove a measurable amount of material from the target surface, resulting in a weight loss from the target. The target is removable so that it may be weighed before and after a sputtering run. The weight loss m divided by the product of the atomic weight M and the mass of a nucleon (1.66×10^{-24} g) gives the number of atoms that have been removed by sputtering. During the sputtering run, the target current has been monitored, and the length of time the sputtering voltage has been applied is recorded. Needless to say, the sputtering voltage, and thus the ion energy, has been held constant during the run, and its value has been recorded. The sputtering current i^+ multiplied by the time t gives the total electric charge carried by the total number of ions incident on the target. Dividing by the charge of an ion e^+ gives the total number of ions incident on the target except for the $(1 + \gamma)$ factor mentioned previously. We actually measure the electron current (i^- rather than the ion current i^+ and recall that i^- includes secondary electrons as well as electrons that neutralize i^+. The relationship is $i^- = i^+(1 + \gamma)$ or $i^+ = i^-/(1 + \gamma)$. Thus the total number of ions incident on the target is given by i^+t/e^+, or $i^-t/e^+(1 + \gamma)$. Since $e^+ = 1.6 \times 10^{-19}$ C, the sputtering yield is given by

$$S = \frac{m/(1.66 \times 10^{-24})M}{i^-t/e^+(1 + \gamma)} = \frac{(9.64 \times 10^4)m\,(1 + \gamma)}{Mti^-}.$$

The secondary emission coefficient γ is not known and is of the order of a few percent so sputtering yields are normally given as $S/(1 + \gamma)$ as in the curves of Figure 7.

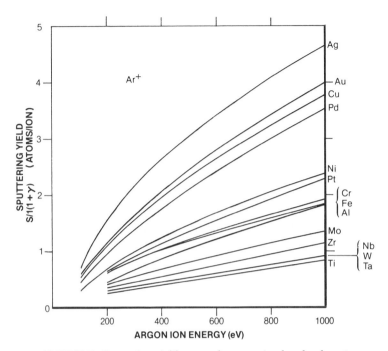

FIGURE 7 Sputtering yield curves for argon ion bombardment.

The interpretation of sputtering yield curves is quite straightforward. We generally treat γ, which is of the order of a few percent at most, as if it were equal to zero. A point on a yield curve is therefore interpreted as being the average number of atoms ejected per ion impact at that ion energy. On the Nb-yield curve we find, at an ion energy of 400 eV, a yield of 0.5; i.e., on the average, half the argon ions striking a Nb surface at 400 eV will cause the ejection of a Nb atom. Clearly, at least half of such impacts and very probably more than half result in no atom ejection at all. It is most probable that there are some impacts that result in the ejection of more than one atom, slightly less than half result in ejection of one atom, and slightly more than half result in no atom ejection at all. Averaging all these events together during a sputtering run results in the total number of Nb atoms ejected being equal to half the total number of incident argon ions at 400 eV. The 400-eV point on the Cr-yield curve is interpreted in a similar way: on the average, one Cr atom is ejected for each 400-eV argon ion impact. Some impacts result in no ejection, some impacts result in ejection of one atom, and some impacts result in the ejection of more than one atom. Averaging all these events together during a sputtering run results in the total number of Cr atoms ejected being equal to the total number of 400-eV argon ions incident on the surface.

Although most sputtering is done using argon ions, it is of interest to compare sputtering yields of various ions as is done in Table I. Here the elements are listed in the order of increasing atomic weights, with the rare gases interposed at the appropriate points for reference. With a few exceptions, none of which appears in this listing, elements listed in the order of increasing atomic weight are in the same sequence as elements listed in the order of increasing atomic number. An interesting correlation that has been noted often in the literature is that sputtering yields at any given ion energy for any given ion vary in a periodic manner when plotted versus atomic number. This is seen in Table I, where for each type of bombarding ion the sputtering yield for Ti is low and the sputtering yield for each successive element increases, reaching a maximum for Cu. The electronic structure of this sequence of elements has one or two electrons in the s level of the N shell, and added electrons enter the d level of the M shell, this level being completely filled at Cu. It is interesting that a secondary maximum occurs at Cr, where the d level is exactly half filled. Going on to higher atomic numbers, the yield for Zr is low and increases to a maximum for Ag. The electronic structure of this sequence of elements has one or two electrons in the s level of the O shell, and added electrons enter the d level of the N shell, this level being completely filled at Ag. It is not noticeable here, but more complete listings when plotted show either a secondary maximum or an inflection point at Mo, where the d level is exactly half filled. The last cycle is from Ta to Au. Here there are one or two electrons in the s level of the P shell, and added electrons enter the d level of the O shell, this level being completely filled at Au. A secondary maximum or inflection point occurs at Re (not listed here), where the d level is exactly half filled.

TABLE I

Sputtering Yields at 600 eV Ion Energy

Element	M	Bombarding ion					
		He^+	Ne^+	Ar^+	Kr^+	Xe^+	Hg^+
He	4.0	—	—	—	—	—	—
Ne	20.2	—	—	—	—	—	—
Al	27.0	0.19	0.85	1.20	1.10	1.00	0.70
Si	28.1	0.16	—	0.50	0.60	0.50	0.20
Ar	39.9	—	—	—	—	—	—
Ti	47.9	0.08	—	0.55	0.55	0.50	0.50
V	50.9	0.07	0.55	0.70	0.70	0.70	0.45
Cr	52.0	0.21	1.00	1.35	1.55	1.90	1.00
Fe	55.8	0.17	0.95	1.30	1.25	1.20	0.85
Ni	58.7	0.18	1.35	1.65	1.55	1.45	1.10
Cu	63.5	0.26	2.00	2.65	2.80	2.45	2.00
Ge	72.6	0.09	0.75	1.25	1.35	1.20	0.90
Kr	83.8	—	—	—	—	—	—
Zr	91.2	0.02	0.40	0.75	0.70	0.70	0.55
Nb	92.9	0.03	0.40	0.60	0.70	0.60	0.50
Mo	95.9	0.04	0.55	0.90	1.05	1.05	0.85
Pd	106.4	0.16	1.30	2.40	2.55	2.50	1.70
Ag	107.9	0.22	2.00	3.40	3.90	3.90	3.00
Xe	131.3	—	—	—	—	—	—
Ta	180.9	0.01	0.30	0.60	0.95	1.00	0.70
W	183.8	0.01	0.35	0.60	1.05	1.15	0.90
Pt	195.1	0.03	0.70	1.55	2.10	2.25	2.60
Au	197.0	0.08	1.15	2.75	3.40	3.40	2.90
Hg	200.6	—	—	—	—	—	—

Considering Table I with the objective of seeing the effect that the type of bombarding ion has on the sputtering yield of a given element, we note that increased difference between the mass of the element and the mass of the bombarding ion results in lower sputtering yield. He^+, for instance, is much lighter than Al, the lightest element for which data are listed here, and the sputtering yield for Al under He^+ bombardment is very low. As the discrepancy between the mass of He^+ and the mass of the other elements listed increases we see, superimposed on the cyclic variation of sputtering yield, a decrease of yield with increased atomic mass. In the case of Xe^+, we have an ion that is much more massive than Al so the sputtering yield for Al under Xe^+ bombardment is lower than under Ar^+ and Kr^+ bombardment. The mass fit with Xe^+ improves with increasing atomic mass so the sputtering yield increases and, as compared with other noble gas ions, becomes greater for elements of mass greater than Xe. Considering Al again, we see that the mass of Al is closer to Ne^+ than to Ar^+ but the yield for Ar^+ is greater than for Ne^+ bombardment. Here we are probably seeing the result of the fact that an atomic particle incident on a surface made up of heavier atoms may strike one of the surface atoms nearly head on and bounce back away from the surface. The rebounding particle in this case imparts some of its original energy to surface atoms, but carries the rest of its original energy away. A particle heavier than any surface atom continues after impact in a direction toward the surface interior regardless of whether the primary collision is head on or not and thus imparts more energy to surface atoms. This would tend to result in a higher probability for the heavier particle to cause ejection of surface atoms. Although the data of Table I are for 600-eV ions, these same general results are found at both lower and higher ion energies.

We shall find that most of our sputtering work is done at ion energies of about 2000 eV so these data do not cover the energy range in which we are most interested. There are data available for the higher energies but not to the extent that has been obtained by Wehner's group at the lower energies. The ratio of the yield for one material to the yield for another material is approximately the same at 600 eV as at 2000 eV so we can make use of these data for first-approximation estimates of yields at higher energies where other data are not available. Most sputtering yields listed in Table II were found in this manner.

It is very useful, for purposes of quick estimates of sputtering rates, to know the sputtering rate at an ion energy at which the sputtering yield is 1 and at an ion current density of 1 mA/cm^2 or 10^{-3} A/cm^2. A current of 1 A corresponds to a charge transfer rate of 1 C/sec so an ion current density of 1 mA/cm^2 corresponds to a charge transfer density of 10^{-3} C/sec cm^2. The charge on a singly charged ion is 1.6×10^{-19} C from which we find that the ion impact density is 6.25×10^{15} ions/sec cm^2. A sputtering yield of

TABLE II[a]

	v_{EV} ($\times$ 10^{-4} cm/sec)	v_{SP} ($\times$ 10^{-5} cm/sec)	Diameter (Å)	S (at 2000 V) (atoms/ion)
Al	11.58	9.33	2.55	2.14
Cr	9.00	7.07	2.29	2.26
Cu	7.85	5.33	2.28	4.66
Au	4.58	4.77	2.57	4.86
Ge	—	—	2.82	2.23
Fe	8.87	7.07	2.28	2.18
Mg	8.33	—	2.85	—
Mo	8.81	6.59	2.50	1.60
Ni	8.74	7.60	2.22	2.91
Nb	—	—	2.62	1.13
Pd	6.51	5.28	2.45	4.18
Pt	5.59	5.60	2.47	2.70
Si	12.60	8.08	2.69	0.92
Ag	5.53	3.95	2.57	6.04
Ta	6.89	6.13	2.63	1.10
Ti	10.17	7.47	2.60	1.01
W	7.08	6.13	2.51	1.13
V	10.26	6.48	2.42	1.22
Zr	8.58	6.75	2.86	1.27
SiO_2	—	—	—	0.38[b]
Stainless steel	—	—	—	1.04

[a]v_{EV} (v_{SP}), ejection velocity of evaporated (sputtered) atoms; diameter, atomic diameter; S, sputtering yield under argon ion bombardment.

[b]Molecules per ion.

one atom ejected per incident ion at an ion current density of 1 mA/cm^2 therefore results in a sputtering rate, i.e., the rate at which atoms are removed from the target surface, of 6.25×10^{15} atoms/sec cm^2. This is not a particularly useful form in which to express the sputtering rate, but we can convert it to two other forms that are useful. In Table II we can see that the average atomic diameter is approximately 2.5 Å (2.5×10^{-8} cm). One atom therefore occupies a surface area of 6.25×10^{-16} cm^2; putting this in another way, a surface area of 1 cm^2 contains 1.6×10^{15} atoms. The density of atoms on a target surface is therefore 1.6×10^{15} atoms/cm^2. At an ion current density of 1 mA/cm^2 and an ion energy at which the sputtering yield is 1, material is removed from the target surface at a rate of 6.25×10^{15} atoms/cm^2 sec, which corresponds to a removal rate of 3.9 atomic layers/sec. It is more convenient to remember this as a rate of 4 atomic layers/sec, and the simplistic assumptions involved in these calculations certainly allow us to consider this to be equally satisfactory. Remembering that the atomic diameter is 2.5 Å, we know that the thickness of an atomic layer is 2.5 Å. A rate of removal of 4 atomic layers/sec therefore corresponds to a removal rate of 10 Å/sec. To reiterate, at an ion energy such that the sputtering yield is 1 atom/ion and at an ion current density of 1 mA/cm^2, the rate of removal of material is 4 atomic layers/sec, or 10 Å/sec. If the sputtering yield is Y and the ion current density is j^+, the rate of removal of material is $4Yj^+$ atomic layers/sec, or $10Yj^+$ Å/sec. These simple expressions are quite useful in making first-approximation estimates in sputtering.

It is a simple matter to convert the system of Figure 6 to that of Figure 8, which is more appropriate for depositing thin film coatings onto substrates. This system is of interest partly because of its use in early sputtering studies, partly because of its use in spectroscopic studies of sputtering, and partly because it has led to the type of sputtering system in use today. The target shield in this system prevents wasted sputtering from the back of the target and serves also to prevent oblique ion incidence at the edge of the target. The situation that would arise at the target edge without shielding is depicted in Figure 9. Over most of the target surface, the ion sheath is parallel to the target surface on both faces, and ions impinge on the target at normal incidence. At the target edge, the electric field approaches that which would exist between con-

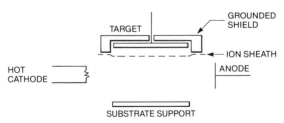

FIGURE 8   Triode sputtering system for thin film deposition.

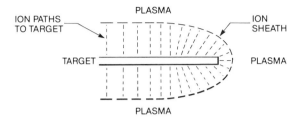

FIGURE 9   Ion sheath at a sharp edge.

centric cylinders, which results in a thinner ion sheath, so the ion sheath takes the shape depicted in Figure 9. This shape, which can be readily observed with a thin, unshielded target, results in a concentration of obliquely incident ions a few millimeters in from the target edge. The combination of concentration plus oblique incidence results in such a high sputtering rate along a line a few millimeters in from the edge that this line sputters through, and an edge strip becomes separated from the main part of the target. If shielding were not used to avoid this problem, the useful life of a target would be greatly shortened.

With some modification to enhance the plasma density and to increase the light emission, the system of Figure 8 was used in spectroscopic studies of sputtering. The light emitted by excited gas atoms in the plasma is emitted at only certain wavelengths characteristic of the plasma gas. When viewed by eye, this light may be blue, red, orange, or some other color, depending on the discharge gas. When viewed in a spectroscope, this light appears as the line spectrum characteristic of the plasma gas, each line corresponding to a different wavelength. These lines extend over the whole spectrum from the ultraviolet through the visible into the infrared. Some lines are more intense than others, and the most intense visible lines determine the visual color of the discharge. If, for instance, the blue lines predominate, then the appearance of the light from the plasma will be mostly blue to the unaided eye. When atoms are sputtered from the target, they are exposed to the same ionization and excitation conditions as the plasma gas atoms, and they emit their characteristic spectrum. This is visible to the unaided eye as a change in color of the plasma at the instant sputtering voltage is applied to the target. When viewed with a spectroscope, this is visible as the line spectrum of the target atoms superimposed on the line spectrum of the plasma gas. A monochrometer can be used to single out a strong emission line of the target so that its intensity can be measured with a photomultiplier. We can be sure that the greater the density of target atoms in the discharge, the higher the intensity of light emitted by target atoms. We know that the higher the sputtering yield, the greater the density of target atoms in the discharge. It is not unreasonable to assume that the spectral line intensity is proportional to the sputtering yield. We can obtain spectral line intensity data as a function of bombarding ion energy over the same range of energies covered by the gravimetric method and down to energies much below the minimum energies at which the gravimetric method is reliable. We can obtain a calibration factor for a given material by dividing the measured sputtering yield of that material at a given energy by the measured intensity of a spectral line of that material at that energy. If this calibration factor is constant, then spectral line intensity is indeed proportional to sputtering yield. In each case, the calibration factor is found to be constant over the range of ion energies covered by both methods, establishing that we are justified in assuming direct proportionality between sputtering yield and spectral line intensity. Using this constant of proportionality, we can use the spectroscopic data to extend sputtering yield curves to lower energies, where we find that the results strongly suggest that each material has a threshold below which sputtering does not occur.

A variation of the spectroscopic method was used to determine velocities of sputtered atoms. Masking and a small observation port were used to limit the light entering the monochromator to that emitted within a small observation volume at a known distance away from the target. The target was pulsed to a negative voltage for 1 μsec so that a small group of atoms was sputtered from the target, essentially at one time. The light emitted by atoms in this group as they passed through the observation volume was detected by the photomultiplier. The time interval between the application of the voltage pulse on the target and the detection of the atoms in the observation volume is the time of flight. The flight distance divided by the time of flight is the velocity. It was found that, just as in the case of evaporated atoms, there is a distribution of velocities of sputtered atoms. Some atoms are ejected with very low energies (near zero) and some with energies nearly an order of magnitude greater than the average. There are major differences from the case of evaporated atoms. Sputtered atoms have average velocities (listed in Table I under v_{SP}) that do not vary a great deal from one type of material to another, being in the range 4 to 8×10^5 cm/sec for most materials. To compare these velocities with velocities of evaporated atoms, we have used our earlier expression $v = 1.58 \times 10^4 \sqrt{T/M}$ to calculate the average velocities of evaporated atoms at T_{EV} and listed these values as v_{EV} in Table II. We see that average velocities of sputtered atoms are an order of magnitude greater than average velocities of evaporated atoms. Since energy is proportional to the square of the velocity, average energies of sputtered atoms are two orders of magnitude greater than average energies of evaporated atoms. Some sputtered atoms have energies high enough to cause sputtering when they strike walls, substrates, fixturing, etc., but these energies are in the low yield range and only a negligibly small fraction of sputtered atoms have these energies.

Radio-Frequency Sputtering

As sputtering developed into a science and a reliable technology, applications arose in which sputtering of nonconductors would be useful. No sputtering system could be used for depositing nonconductors because it was always necessary to apply a negative voltage to the target, and this can be done only with conductive targets. If we now consider the target arrangement of Figure 10, we see how it is possible to sputter nonconductive materials. If a conductive plate is placed behind a nonconductive plate in a plasma, then a negative voltage may be applied to the conductive plate. Ions will be accelerated toward the conductive plate, but they will impinge on the intervening nonconductive plate and cause sputtering of the nonconductor. This bombardment will last for about 10^{-7} sec, after which time there will have been built up a positive

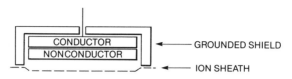

FIGURE 10   Radio-frequency sputtering target for nonconductors.

charge on the surface of the insulator that will counteract the negative potential applied to the conductor. This terminates the high
energy ion bombardment and sputtering after 10^{-7} sec. Even
though we already realize that this is not a practical method of
sputtering, we shall pursue this course of action further. If we now
reverse the connections to the power supply and apply a positive
potential to the conductor, electrons will be accelerated toward the
conductor, strike the intervening insulator, and build up a negative
surface charge (in about 10^{-9} sec) that will counteract the positive
potential. We are then in position to again reverse the connections
to the power supply and achieve another 10^{-7}-sec burst of sputtering of the insulator. Every other reversal of the connections will
give 10^{-7} sec of sputtering, but we need an absolute minimum of
100-sec sputtering time. To achieve this, we would have to make
10^9 double reversals of the connections (as an absolute minimum),
which would be a tiresome and time-consuming task. Our first
thought may then be that we can accomplish this by connecting to
an ac transformer plugged into a 60-Hz (hertz or cycles per second)
wall outlet. To get 10^9 double reversals by this means will take
$10^9/60$, or 1.67×10^7, sec (193 days), which is not very practical
either. We are on the right track though, and, if we use a radio-
frequency (rf) generator operating at a frequency of a megacycle
per second or more, the rate of reversal is high enough that we can
sputter insulators at a practical rate. This idea was first suggested
by Wehner at about the time that he was putting sputtering on a
real scientific basis. There was such a great amount of basic work to
be done at that time that it was about ten years before Wehner's
group found time to act on the suggestion and develop the first rf
sputtering system.

   The first rf sputtering systems were simply the dc system of
Figure 8 with the dc target replaced by the rf target of Figure 10 and

with an rf generator and matching network replacing the dc target power supply. Unlike the dc power supply, which was usually referenced to the anode as in Figure 6, the rf power supply was referenced to ground. It was soon found that the anode–hot-cathode system for generating plasma was superfluous. It was found that the rf power alone applied to the target could both generate plasma and accelerate ions to the target to cause sputtering. If the rf power was turned on before the rest of the system, it generated a plasma and accelerated ions to the target to cause sputtering. If the anode–hot-cathode system was generating a plasma and malfunctioned during the course of an rf sputtering run, the rf continued to generate a plasma and continued to sputter the target. There is a difference in tuning when the rf generator is providing power for both sputtering and plasma generation, but otherwise sputtering is the same in either case.

There is, of course, no reason that rf should not be used to sputter conductors as well as insulators. The only additional need for sputtering conductors is for a blocking capacitor to prevent shorting the target to ground through the matching network. The rf sputtering system thus becomes a simple, effective, all-purpose sputtering system.

A typical rf sputtering system is depicted in Figure 11. The gas (argon) inlet and vacuum pumping system are not shown here. The forward- and reflected-power meters shown separately here are commonly incorporated into the rf generator chassis. Radio-frequency generators are designed to have an output impedance of

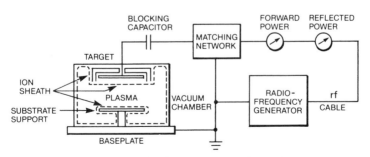

FIGURE 11   Radio-frequency sputtering system.

50 Ω to match the available transmission cable and must feed into a load impedance of 50 Ω. The actual load impedance, the target, is effectively a capacitor with a bypass resistance quite a bit different from 50 Ω. The matching network together with the capacitance of the target forms a resonant circuit so that the rf power is dissipated at the resistance. The matching network also acts like a transformer to adjust the actual resistance to the 50-Ω value needed to match the transmission cable and the rf generator. The target, together with the baseplate and other grounded portions of the chamber, forms a simple diode (two electrode) sputtering system in much the same manner as the original dc glow discharge sputtering system. Unlike the dc glow discharge system, the rf system can operate at the low gas pressures needed for reliable sputtering work. Unlike the dc glow discharge system, the rf system is capable of relatively high coating rates. Unlike the dc glow discharge system, the rf system can deposit both conductive materials and insulating materials.

Radio-frequency sputtering is a very simple operation. The proper target is installed; a test piece is cleaned, weighed, and installed; and the chamber is closed and evacuated to a good vacuum. The argon pressure is then adjusted, and the rf generator is turned on and set for about one-fourth power. The two adjustment controls of the matching network are then tuned to give zero reflected power. Subsequent steps of increasing the power level and retuning the matching network are needed to bring the power up to the desired operating level. The substrates are covered by a shield (not shown in Figure 11) during pumpdown, during initial tuning of the matching network, and during sputter cleaning of the target. A few minutes at the desired operating power level will sputter the target surface clean and ensure that the sputtered deposit will be pure. The shield is then opened, and the test piece is coated long enough to increase its weight by a significantly measurable amount. The coating time τ and power level together with all standard operating parameters are recorded. The test piece is removed from the vacuum system and weighed. The weight gain m together with the known density d of the material and the known surface area A of the test piece allows the calculation of the thickness t of the deposit from $m = dAt$. The coating rate $\dot{t}$ is found from t and the known coating time τ by $t = t/\tau$. We thus have a calibration of the sputtering system that provides a very simple thickness control monitor. We simply calculate the coating time required to give the desired coating thickness and coat for that length of time. It is useful to include a test piece with each coating run to verify that the proper thickness has been deposited and that the calibration is remaining constant. After the initial calibration run the matching network needs only minor retuning as long as the target is not changed. It is necessary to shield the substrates and sputter the target clean each time new substrates are installed.

The sputter cleaning of the target serves two additional useful functions. The plasma always assumes a potential somewhat positive with respect to all surfaces to which it is exposed. As a consequence, all interior surfaces of the sputtering system are subjected to ion bombardment. The energies of ions impinging on surfaces other than the target are too low to cause sputtering, but are high enough to be very effective in causing desorption of gases. This scrubbing action is so effective that a burst of gas is detected (as an increase of pressure in the foreline) at the time the discharge starts. The foreline pressure goes back down to its equilibrium value after a moment or two. In addition to the scrubbing action of the discharge, there is a highly effective pumping action due to the sputtering of material from the target. The material sputtered during the target cleanup period will deposit on exposed surfaces including the back of the shield that is protecting the substrates. This deposit, being atomically clean, is highly reactive and combines strongly with any reactive gas remaining in the chamber. This action, known as gettering, removes reactive (contaminating) gases from the chamber at a high rate, i.e., provides a very efficient pumping action. As a consequence, the contaminant gas levels in the coating chamber are far below the 10^{-7} Torr range indicated by the ion gauge prior to the start of the sputtering operation. Between the scrubbing action and the gettering action, there is good reason to believe that the contaminant level in the chamber during sputtering remains below 10^{-9} Torr and may be as low as 10^{-12} Torr.

Radio-frequency sputtering is very useful as a means of depositing thin film coatings onto parts and substrates. The rf system is very simple to operate and gives reliable, repeatable coatings. As a tool for studies of sputtering, rf systems are not satisfactory because the energies of the impinging ions are not well defined. When sputtering from a conductive target in an rf system, one can connect a dc voltmeter to the target via a suitable blocking inductance and measure an average dc voltage. This voltage depends on the rf power, the size and shape of the target, and the type of gas, but is generally in the range 1000–2000 V. It is not certain what type of an average this is, but it is certain that ion energies range above and below this average and are therefore not well defined. This is of no consequence in depositing thin films because the only requirements here are repeatability and reliability, and rf sputtering systems meet these requirements in the most satisfactory manner.

We have not spoken of secondary electrons in sputtering other than with respect to their influence on sputtering yield measurements. Secondary electrons can have a strong influence on coating results as well. Secondary electrons released at the target surface are accelerated by the same mechanisms that result in an average target voltage of 1000 to 2000 V and acquire energies of this average value. Since these electrons are accelerated directly away from the target, they are accelerated directly toward the substrates and give up their energy to the substrates upon impact. The resulting heating of the substrates may be detrimental to the substrates or to the deposit or to both. The secondary electrons can be prevented from striking the substrates by using a transverse magnetic field to deflect them, and this is fairly standard practice in sputtering. The magnetic field also increases the sputtering efficiency by increasing the plasma density.

The scrubbing action of the low energy ion bombardment of surfaces exposed to a plasma has been discussed previously. Low energy ions are very effective in causing desorption of gases from surfaces. We have often spoken of the high reactivity of atomically clean metallic surfaces, and it is clear from this that atomically clean surfaces exist only under vacuum or pure inert gas environments. We have also indicated previously that all surfaces adsorb gases and that all gases are adsorbed to some extent by any surface. Any substrate installed for coating thus has two types of surface contamination: weakly bound, adsorbed gases and strongly bound, chemically bonded elements. The adsorbed gases can be removed quite well simply by exposure to the plasma, and in many cases this is adequate. In many other cases it is desirable to remove the chemically bonded elements, and sputtering provides a unique opportunity to remove these elements. We need only apply to the substrates a negative voltage of a value high enough to cause sputtering of the substrate surface. This is spoken of as reverse sputtering, backsputtering, or sputter-cleaning. Radio-frequency power is applied to the target to generate a plasma and to sputter the target, and at the same time dc power is applied to the metallic substrate to sputter-clean the substrate. Some material thus must be sputtered back and forth between target and substrate, but there are enough losses to open areas within the chamber that the substrate surface will be atomically clean within a few minutes. Many materials that do not adhere well to normal surfaces will bond strongly to such atomically clean surfaces. Copper, silver, and gold are typical examples of such materials.

In many cases it is not feasible to sputter-clean substrates, but the desired coating material will not adhere well to anything except an atomically clean surface. These substrates may be coated by use of an intermediate coating, often referred to as a bonding layer. A thin coating of a material that is known to adhere well to the substrate is deposited first. The surface of a freshly deposited coating is atomically clean and, in a good vacuum, will remain atomically clean for some time. The desired coating material is then deposited onto this atomically clean surface, to which it will adhere strongly. It is important that the intermediate coating not be exposed to air or to a poor vacuum prior to depositing the second coating. It is similarly important to minimize the time the atomically clean surface remains exposed to even a very good vacuum before the second coating is deposited. It is usually feasible to start the second coating within 10 sec of the time that the initial coating is completed. We have had delays as long as 10 min between coatings with no detectable loss of adhesive strength, but it is preferable to minimize the delay between coatings. We have made no effort to determine the maximum permissible time delay between coatings.

We recall that we had calculated that a surface atom in the presence of gas at 1-mTorr pressure experiences 267 collisions/sec. If this is a reactive gas, the lifetime of a clean surface at 1 mTorr (10^{-3} Torr) is 1/267 sec, or 3.7×10^{-3} sec. We know that the number of collisions per second is directly proportional to pressure so the lifetime of a clean surface is inversely proportional to pressure. Therefore, at a nominal chamber pressure of 10^{-7} Torr of contaminant gases, the lifetime of an atomically clean surface is 3.7×10^{-3} sec $\times$ (10^{-3} Torr/10^{-7} Torr), or 37 sec. If the ion gauge is correctly reading the residual gas level and the residual gases are contaminant gases, an atomically clean surface can be maintained for only 37 sec in vacuum systems of the type discussed in this book. Actually, in 37 sec the surface would be 100% contaminated so the atomically clean condition would exist for less than a second. Nevertheless, we know that atomically clean surfaces have been maintained for as long as 10 min (600 sec) in these systems, so the sputtering chamber contaminant level must be below 6×10^{-9} Torr. We are convinced that contaminant levels during sputtering are appreciably below this, possibly as low as 10^{-12} Torr.

Unique Characteristics of Sputtering

It is not uncommon to try to make comparisons of the evaporation and the sputtering methods of depositing thin film coatings, looking for differences in quality, but it quickly becomes apparent that this is not a useful approach. The choice between evaporation and sputtering is usually a matter of convenience, although there are some things that can be done only by sputtering. Some of the unique characteristics of sputtering that are considered in evaluation of this coating method are the following.

1. Deposition rates do not differ a great deal from one material to another. This is often a useful feature in multilayer depositions.

2. Deposition rates by sputtering are much lower for some materials and much higher for other materials than deposition rates by evaporation.

3. Thickness control is very simple. After a calibration run has been made, thickness control is merely a matter of setting a timer.

4. The lifetime of a sputtering target may be as long as hundreds of runs and is seldom less than 20. This is in sharp contrast to evaporation, where a source seldom lasts as long as 10 runs.

5. In sputtering alloys and other complicated materials, the deposit maintains stoichiometry with the original target composition. No other method can be used to deposit alloys.

6. Cleaning of parts and substrates by reverse sputtering is an advantage that can be gained with no other process. One can combine sputter-cleaning with vacuum evaporation, but this obviously tends to complicate operations.

7. In Chapter III we discussed the problem of the ejection of particles from sources during evaporation. This problem, known as spitting, does not occur in sputtering.

8. The high ejection energy of sputtered atoms is often suggested as a factor in improving film structure and adhesion to the substrate. This is not very likely to be a detectable factor. It is our experience that, where sputter-cleaning is not used, the adhesion of a given material to a given substrate is the same for sputtered deposits as for evaporated deposits.

Ion Plating

Ion plating is a process in which a substrate is subjected to ion bombardment both before and during the time it is being coated. The coating process used in ion plating is vacuum evaporation from resistance-heated sources. The vacuum procedures in ion plating are the same as those used in sputtering in that the system is first pumped to a good vacuum and argon gas is then admitted to establish a state of dynamic equilibrium in the chamber. The chamber pressure remains constant with argon being pumped out of the chamber at the same rate as it is flowing into the chamber. Chamber pressures on the order of 40 mTorr are fairly standard in ion plating. A typical ion plating system is depicted in Figure 12. Radio-frequency power is applied to the substrate, as in sputter cleaning, for the purposes of generating a plasma and of causing ion bombardment of the substrate surface. After sputter-cleaning

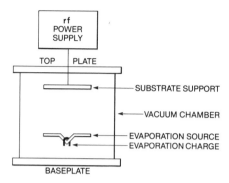

FIGURE 12   Ion plating system.

has removed the contaminants from the substrates to establish atomically clean surfaces, the evaporation source is activated to begin the coating process. At this point ion plating procedures usually are to reduce the rf power so as to limit the rate at which atoms deposited on the substrates are sputtered away. Some rf power is retained on the substrates so that the substrates are subjected to ion bombardment throughout the time coating is being deposited. This ion bombardment consists of argon ions and also ions of the material being deposited on the substrates. It is clear that evaporated atoms, in passing through the plasma on their way from the source to the substrates, are subjected to the same excitation and ionization conditions as argon gas atoms. Some evaporated atoms must therefore become ionized and contribute to the ion bombardment of the substrates.

Among the desirable film properties suggested as resulting from ion plating are film densities approaching bulk material values, strong adhesion to substrates, high deposition rates, high purity of deposited film, uniformity of coating, and the ability to coat the inside of holes and relatively inaccessible hollows. It is a disservice to this excellent vacuum coating process that these suggested properties have been at times presented in a misleading way so as to create expectations that cannot always be fulfilled. We here attempt to discuss these properties to see what can actually be achieved in a practical sense. It is assumed here and throughout these discussions that, regardless of the coating method employed, good clean vacuum conditions are achieved and maintained during a coating run. Nothing deteriorates film properties so much as poor vacuum conditions. This is most severely true in sputtering, but it remains quite severely true in ion plating and vacuum evaporation.

Film densities approaching bulk material values can be achieved using ion plating. The same can be said of sputtered films. The same can be said of evaporated films. Although no truly definitive work seems to have been done in this area, it appears that some material may be quite likely to deposit at densities approaching bulk values regardless of the deposition method. Materials that have a tendency to deposit at densities below bulk values in evaporation work seem to deposit at densities approaching bulk values in sputtering and in ion plating. Sputtered atoms, we recall, arrive at the substrate with energies of the order of 10 eV, a high enough energy to cause disruption of surface atoms, to establish new bonding sites, and to move weakly bonded atoms to stronger bonding locations. In ion plating, ion bombardment of the substrate provides these functions. In evaporation, heating of the substrate before and during deposition is helpful in increasing film density, but it is true that even then film density may be somewhat less than in ion plated or sputtered films.

The need for high film density should not be automatically assumed. In many cases the density of the deposited film may not be of any importance, and one would then not go to added expense to achieve high density.

The strong adhesion of ion plated films can be attributed to the sputter-cleaning of the substrates prior to deposition. The strong bonding to such atomically clean surfaces has been mentioned earlier. It was also pointed out earlier that the deposition method seemed to have no real influence on the adhesion characteristics between a thin film and an atomically clean surface. We also want to remember that some materials have strong adhesion characteristics when deposited onto almost any reasonably clean, not necessarily atomically clean, surface.

Since the deposition method in ion plating is by vacuum

evaporation, the earlier comments on deposition rates apply here. It is possible to use E-beam deposition in ion plating and achieve these high deposition rates if differential pumping is employed to keep the electron source at suitably low pressure while the rest of the chamber is exposed to relatively high argon pressure. Earlier discussions of film purity are also applicable here. It is at times implied that the ion bombardment during deposition in ion plating may help to maintain surface cleanliness and film purity, but this seems unlikely. We know that the gettering action of freshly depositing material is remarkably effective in maintaining extremely clean vacuum conditions. It is therefore improbable that the surfaces being coated would suffer contamination in any way that ion bombardment could alleviate.

Coating uniformity was discussed earlier, and it is clear that the same considerations are applicable in the case of ion plating. Planetary fixturing, which helps to achieve uniformity in sputtering and vacuum evaporation, is not practical in ion plating.

With ion plating there is a somewhat improved capability of coating inside of holes in some cases, but this is usually not greatly important. The coating inside the holes does not have the strong adhesion found with atomically clean surfaces, but this is no detraction because the coating here is usually not subjected to stress. The mechanism by which ion plating is better able to deposit atoms in the interior of holes is occasionally said to be ionized atoms following electric field lines around corners and into holes. There are several reasons that this mechanism is not likely to be significant. We know that electrons do not follow field lines that closely so we can be quite sure that particles as heavy as ions do not do so. Even to the limited extent that ions may follow field lines, there are not that many field lines terminating inside of holes. Beyond this, the percentage of evaporated atoms ionized in passing through the plasma is quite low, thus putting another limitation on the efficiency of this mechanism. A more likely mechanism for projecting coating into holes is the ion bombardment during deposition in ion plating. Material directly deposited in a hole is deposited mostly just inside the hole entrance. Ions striking here are, of course, incident at oblique angles resulting in forward sputtering of this deposit deeper into the hole.

Ion plating, like sputtering and vacuum evaporation, has its more useful applications. One must make certain to analyze the application requirements correctly so as to select the most suitable coating method. One must also correctly ascertain the true capabilities of each coating method to avoid unrealistic expectations. As a general rule, most thin film applications can be handled acceptably by any one of the three vacuum coating methods.

CHAPTER V

THIN FILMS

Our discussions up to this point have centered on methods of depositing thin films. We want now to make some mention of the uses or applications of thin films. Where we have mentioned thin films previously, it has been implicitly assumed that the object being coated was flat, being coated on one surface only, and being coated over the entire flat surface. Sputtering and evaporation are line-of-sight coating methods. If a three-dimensional object is to be coated over all surfaces, rotation and/or multiple sources must be employed. In many applications the requirement is to coat only certain specified areas of the substrate. Where only the crudest definition of the edges of the coating is acceptable, aluminum foil can be wrapped around areas or folded into masks to shield areas that are to be kept free of coating. Better definition of the edges of the coated area can be achieved by machining masks out of sheet metal. Somewhat closer tolerances can be met by using pho-tolithographic methods to etch masks from metal foil. The greatest precision is achieved by coating the entire surface of a substrate and then employing photolithographic methods to remove the un-wanted coating and leave the desired thin film pattern on the substrate surface.

Photolithography

Photolithography makes use of resins that become photosensitive when applied to a surface and allowed to dry. These resins, which we commonly speak of as photoresists, are available dissolved in solvents and are not photosensitive in this form. They may be applied by brushing, rolling, spraying, dipping, etc., or they may be poured onto a surface that is then spun at some rate in the range 50–5000 RPM to spread the photoresist into a thin uniform coating. After drying, these coatings are photosensitive, principally to ultraviolet radiation. Some exposure to normal room light is usually not detrimental to the Kodak KPR series of photoresists or to the Kodak KTFR photoresist.

Figure 1 depicts a sequence in which a thin film has been deposited onto a substrate and photolithographic techniques are being employed to remove unwanted coating and leave a simple thin film pattern on the substrate. The top frame in the sequence shows the photoresist being exposed to ultraviolet light through a photographic mask of the desired thin film pattern. In this mask the image areas where thin film material is to be etched away show up as opaque areas and the image areas where thin film coating is to remain on the substrate show up as transparent areas. It is important to be aware that photographic masks are composed of a relatively thick transparent backing to support the thin emulsion that contains the pattern. The pattern, i.e., the emulsion side of the msk, must be in intimate contact with the photoresist surface to achieve sharp definition. Vacuum hold down is essential to ensure that contact between the mask and the photoresist surface is intimate over the entire surface. Loose contact would allow light to scatter under the opaque areas and expose photoresist that is not intended to be exposed.

The second frame of Figure 1 shows the substrate after the photoresist has been exposed and developed. The areas of photoresist that are subjected to exposure to ultraviolet light become insoluble in the developer solvent, whereas the unexposed areas are dissolved by the developer solvent. After exposure and development, the exposed areas of photoresist therefore remain bonded to the thin film, protecting the areas where the thin film coating is to

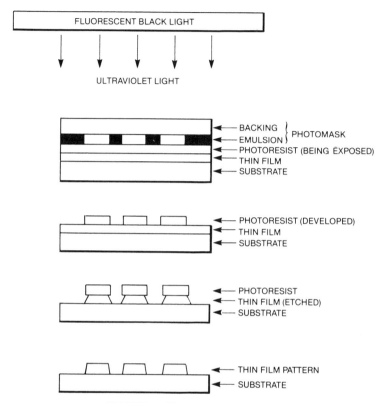

FLUORESCENT BLACK LIGHT

ULTRAVIOLET LIGHT

BACKING
EMULSION } PHOTOMASK
PHOTORESIST (BEING EXPOSED)
THIN FILM
SUBSTRATE

PHOTORESIST (DEVELOPED)
THIN FILM
SUBSTRATE

PHOTORESIST
THIN FILM (ETCHED)
SUBSTRATE

THIN FILM PATTERN
SUBSTRATE

FIGURE 1 Photoetching sequence.

remain on the substrate. The unexposed areas of photoresist are dissolved away during the developing process, thus exposing the areas where the thin film coating is to be etched away.

The third frame of Figure 1 shows the substrate after it has been processed in an etching solution. The unprotected thin film has been etched away, and the thin film that has been protected by photoresist remains on the substrate. The frame shows undercutting where, as might be expected, the etchant eats into the thin film underneath the edges of the protective photoresist. This is usually not a serious problem in this film work because the amount of undercutting is, in the worst case, only about 25% of the thickness of the thin film. Line definition specifications usually allow much greater variation than even the total thickness of a thin film. The fourth frame of Figure 1 is quite trivial, showing the etched pattern after the photoresist has been removed. As a rule, photoresist is most conveniently removed by use of photoresist thinner.

This brief description of photolighography has considered only the types known as negative-acting photoresists. Negative-acting photoresists are soluble in the developer unless they have been exposed to ultraviolet light. Exposure to ultraviolet light causes negative-acting photoresists to become insoluble in the developer. There are available positive-acting photoresists that are insoluble in the developer until after they have been exposed to ultraviolet light. An excellent discussion of photolithographic techniques is found in Kodak pamphlet P-246 entitled Photofabrication Methods. This pamphlet may not make it entirely clear that the vacuum hold down may be extremely simplistic and still be most satisfactory. It may also not make it clear that the ultraviolet light source may be simply a fluorescent black light of the type available in almost any lighting fixture store. One may finally get the impression from this pamphlet that one should prepare one's own photomasks, but, in fact, it is generally best to have them prepared by a company specializing in this service. Companies providing this service may be listed in the business telephone directory under Printed and Etched Circuits. Many companies listed under this heading provide the entire service of preparing the photomask and carrying out the photoetching process.

General Applications

The applications of thin film coatings include coatings for decorative purposes, optical functions, optical patterns, electrical insulation, electrical conductivity, resistance networks, corrosion resistance, wear resistance, solid lubricants, brazing, and braze diffusion bonding. Probably the most widespread use of vacuum deposited thin films is in the decorative coating of plastic and cast toys, trophies, automotive interior trim items, and cosmetic bottles. The reflective coating on most of these objects is a vacuum evaporated coating of aluminum about 1500 Å thick sandwiched between two layers of clear lacquer. The part is coated with clear lacquer to serve as a base for the aluminum coating, oven baked, vacuum coated with aluminum, coated again with clear lacquer to serve as a protective layer, and oven baked again. The result is a beautiful, shiny metallic finish that looks like silver where one might expect silver, as on a trophy, and looks like chrome where one might expect chrome, as on an automotive interior trim part. A part coated in this manner can be given a beautiful, gold tone by soaking it for a few minutes in a dye solution after the final bake.

Another extensive application of vacuum deposited thin films is for antireflection coatings on optical components. Eyeglass lenses, camber lenses, microscope components, binocular components, etc., may be made quite free of reflection problems by a vacuum evaporated coating of magnesium fluoride of thickness of about 1200 Å. At this thickness, interference between light waves reflected from the glass surface and from the magnesium fluoride surface cancels about 95% of the total reflection. This is very helpful in single component optical parts and is nearly indispensable in multiple-lens elements.

Electronic applications may be more widely recognized as applications of vacuum deposited thin films. There are thin film resistors, capacitors, magnetic devices, and active elements, but the most extensive use of vacuum deposited thin films in electronics is as electrical connection elements in integrated circuits. Basically, these electrical connection elements constitute somewhat exotic circuit boards. In some cases these may involve multilayer coatings with conductive layers separated by nonconductive layers. There are cases where one short length of conductive element is caused to bridge over another conductive element. This is accomplished by depositing an easily etched layer between the two conductive layers and etching the intermediate coating away at the bridge point during the photolithographic processing.

Where electrically insulating coatings such as SiO, SiO_2 or Al_2O_3 are used to separate electrically conductive layers, the electrically conductive layers cannot be maintained at more than about a 30-V difference in potential from each other because of electrical breakdown (arcing). This is because no surface can be polished perfectly smooth and flat, and, in fact, the most highly polished surface can be quite rough when compared with the thickness of a thin film. It is certain that there are areas where the microsurface is tilted at large angles with respect to the macrosurface. The film thickness at these areas, being proportional to the cosine of the tilt angle, is much thinner than the overall average and therefore more susceptible to electrical breakdown. The edges of these areas are very likely to have sharp points or edges that will concentrate the electric field and thus increase the susceptibility to electrical breakdown. At first one tends to think of projecting micropoints as being the main factor here, but further consideration leads instead to the belief that the major problem arises from microholes, or voids, in the surface.

The situation where two separate surfaces are involved is quite different. If one conductive surface is coated with an electrically insulating thin film and placed in direct contact with another conductive surface, the thin film can protect against electrical breakdown even with as much as several hundred volts applied. This is because the closest separation between electrically conductive surfaces in this case is governed by the thickest areas rather than the thinnest areas of insulating thin film.

Resistive coatings are usually deposited by sputtering because they are usually alloys, such as Nichrome,* or mixtures, such as cermets, which cannot be deposited reliably by evaporation methods. Cermet-resistive coatings can be deposited by the simultaneous sputtering of a nonmetallic mineral (ceramic) and a metal. The term cermet is obtained by combining cer from ceramic with met from metal. A convenient ceramic for codeposition is fused quartz, and a convenient metal is nickel. A wide range of resistances can be obtained by varying the ratio of metal to ceramic.

Corrosion-resistant coatings are selected according to the environmental elements against which protection is needed. Some coatings are corrosion resistant in many environments, but no coating is corrosion resistant in all environments. Careful consideration of the compatibility between substrate and coating and careful consideration of the potential interaction of the coating with

*Trademark; Driver-Harris Company.

corrosive elements are needed if success is to be achieved in this area.

Wear-resistant and solid-lubricant coatings approach the same problem from opposite directions. At times one wearing surface may be coated with a wear-resistant material while its mating surface is coated with a solid lubricant. Carbides, especially tungsten carbide, are favored wear-resistant coatings, but there are indications that silicon nitride may also have some desirable properties. Molybdenum disulfide and silver are preferred solid lubricants.

Braze materials are most commonly alloys and can be vacuum deposited only by sputtering. Vacuum deposition of braze materials is used when it is important to minimize the amount of braze in the finished part. Another value of vacuum depositing braze material is that the surfaces that are to be brazed together may be sputter-cleaned before being coated. This ensures the cleanest and strongest possible braze joint. Braze diffusion bonding alloys, like regular braze material, can be deposited only by sputtering. This differs from ordinary brazing in that the braze material is of nearly the same alloy composition as the parts being brazed together, but contains a small percentage of a melting-point depressant, such as boron. Parts being joined together by braze diffusion bonding are brought up to brazing temperature and held there to allow the melting-point depressant to diffuse away from the joint region. After diffusion is complete, the percentage of depressant material is at a value low enough that it no longer affects the melting point, and the joint is as perfect as the bulk material. This technique permits the incorporation of highly intricate channels within the body of a solid part. When such a part is to be made up of a number of thin sections, the bonding layers must be very thin to minimize the total amount of melting point depressant in the finished part. Depositing the bonding layers by sputtering provides excellent control.

Specific Applications

An interesting application of thin films is found in Report #AFML-TR-4093 from USAF Wright-Patterson Air Force Base. The purpose of this study had been to investigate the possibility of using the hot isostatic pressure (HIP) process to repair fatigue cracks in turbine wheels. The concept was to vacuum seal the cracks (which have widths on the order of 10^{-3} cm or less) and subject the part to high pressure argon at high temperature. If the cracks are sealed under vacuum conditions, then the potential exists for the hot isostatic pressure to close these cracks and for high temperature diffusion processes to bond the damaged areas back together. Sputtering (with initial sputter-cleaning of the part) was employed to seal the cracks under vacuum conditions. Power was applied simultaneously to both the part and the target, and, when sputter cleaning was complete, coating was instituted by reducing power input to the part and increasing power input to the target. Since a small negative voltage was maintained on the part throughout the coating time, the coating process can reasonably be called ion plating and is thus called in the report. A coating thickness of 3 $\times$ 10^{-3} cm bridged and sealed the cracks. The HIP process was successful in repairing the fatigue cracks, but the parts did not retain dimensional specifications so the concept had to be abandoned.

One wonders if such cracks could have been sealed simply by ion bombardment. If we consider the simplified sketch in Figure 2, we can see how the high sputtering yield and forward ejection pattern resulting from oblique ion incidence conceivably could seal a small crack. Ions incident on undamaged areas of the surface

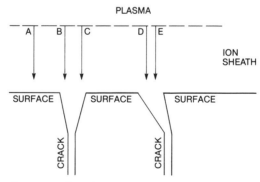

FIGURE 2 Depiction of ion bombardment of the surface of a part which has suffered fatigue cracks.

such as is depicted in event A result in surface erosion with no contribution to the sealing process. Ions incident on damaged areas, such as depicted in events B, C, D, and E, would strike interior portions of cracks obliquely. We recall that oblique incidence results in greatly enhanced sputtering yields and that these sputtered atoms are ejected predominantly in the forward direction. The forward-ejected atoms would pile up at some point inside the crack, and one could anticipate that this might result in closure of the fissures before surface erosion would become excessive.

The Physical Electronics Division of Perkin-Elmer makes use of vacuum deposited thin films in a number of applications in their instruments. Most of these coatings, although quite ingenious, reduce essentially to standard practice; however, several cases are unique. The combined resistive–conductive pattern depicted in Figure 3 is protected by patent. One application of this type of pattern is to generate electric fields essentially for the purpose of focusing electrons. The narrow conductive bands in the interior regions of the pattern enforce equipotential lines at these locations, regardless of possible inhomogenetics in the resistive coating. Resistance from band to band may be adjusted so as to achieve the desired distribution of the applied voltage. The resistive coating is a cermet resulting from codeposition of a ceramic and a metal in a ratio designed to achieve the desired resistivity.

The Physical Electronics Division x-ray anode is a very logical application of vacuum deposited coatings. The x-ray anode is

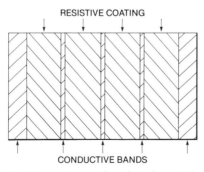

FIGURE 3 Resistive pattern with enforced equipotential bands.

operated at power input levels so high that water cooling is required, but the anode materials are either too expensive to be directly machined into a water-cooled configuration or are subject to corrosion when exposed to water. These problems are circumvented by machining the anode body from OFHC copper and vacuum depositing the desired x-ray source material onto the anode surface. The interior of the OFHC copper anode body is water cooled, and the OFHC copper walls conduct heat away from the x-ray source material more effectively than if the entire body were of the x-ray source material.

A gold coating on aluminum surfaces in Auger Electron Analyzers, manufactured by the Physical Electronics Division of Perkin-Elmer, provides a simple solution to a complex problem. Throughout this book we have encountered the fact that metallic surfaces are so highly reactive that they are unable to remain atomically clean. A metallic surface that might be atomically clean can remain so for only about 5×10^{-9} sec when exposed to air. Within this time, oxidation and nitridation have completely covered the surface and started working deeper into the surface. Aluminum is particularly reactive and therefore particularly susceptible to this problem. These surface layers of oxide and nitride are not thick enough to influence ordinary usage, but they are insulated areas and can become weakly charged by stray electric charges. In the auger electron analyzer, this can distort electric fields so as to cause some defocusing with consequent loss of sensitivity. Since gold is highly resistant to formation of these insulating layers, a gold coating on the aluminum surfaces effectively alleviates these problems at a reasonable cost.

The 3M Company employs vacuum deposited gold films in a totally different application, namely, as mercury-vapor sensors. The affinity of gold for mercury is well known, probably because mercury, through this affinity, has damaged so much gold jewelry. 3M has found that vacuum deposited gold films exposed to mercury vapor undergo permanent changes that are proportional to the amount of exposure. These gold films therefore serve as excellent mercury vapor sensors providing a quantitative means of monitoring an individual's exposure to mercury vapor. Other mercury sensors serve to monitor instantaneous levels of mercury vapor,

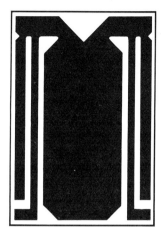

FIGURE 4   Krak-Gage.

but are unable to provide the quantitative measure of total exposure provided by the 3M mercury monitor.

The newly developed version of the Krak-Gage* provides an excellent illustration of a type of application in which it is necessary to go to the added expense of vacuum deposition. The Krak-Gage is a transducer used to monitor the propagation of fatigue cracks in specimens under stress. In standard form (Figure 4) the Krak-Gage is formed by photolithographic processes from resistive sheet stock bonded to an insulating film. The insulating film provides electrical isolation between the Krak-Gage and the specimen (usually metallic) to which it is bonded by an adhesive. The need to conduct fatigue tests under environmental conditions too severe for available bonding adhesives prompted the Hartrun Corporation to develop a vacuum deposited version. In this version, the specimen is first coated with SiO_2 to provide electrical isolation of the Krak-Gage from the test specimen. A resistive film of the same material and the same thickness as the standard Krak-Gage sheet stock is deposited onto the SiO_2 film. Photolithographic techniques are then used to form the Krak-Gage. The bonding is superior to adhesive

*Trademark; Hartrun Corporation.

bonding under ordinary conditions and maintains its integrity under environmental conditions too severe for adhesive bonding.

The Sundstrand Corporation has employed vacuum deposited aluminum oxide coatings in a most useful way in heat treating, brazing, and diffusion bonding operations. A problem often encountered is that metallic fixturing used in vacuum for these high temperature operations may experience diffusion bonding such that the fixturing cannot be disassembled. In particular, threaded parts bonding in place frequently hamper design of fixturing for vacuum furnace operations. Aluminum oxide coatings on machine screws, on mating surfaces, or on any area where contact between metallic objects may occur is highly effective in preventing unwanted bonding.

The discussion of thin films here has been very limited, principally because there is a continuing series of books that provides excellent coverage of this field: "Physics of Thin Films" (Academic Press, New York). Beyond this, new applications are left to the ingenuity of the reader. New applications are continually being discovered, often (possibly most often) by people with no background or experience in thin film technology.

INDEX

Resistivity, 85
Reverse sputtering, 129
rf generator, *see* Radio-frequency
 generator
rf sputtering, *see* Radio-frequency
 sputtering

S

Secondary emission, 95, 99, 114
Source, evaporation, 75
Spectroscopic study, 121
Sputter-cleaning, 127, 129
Sputtering, 91, 100
Sputtering rate, 118
Sputtering yield, 95, 120
Sputtering yield curve, 115
State of matter, 2
Sublimation, 4
Substrate, 87

T

Tape test, 87
Target, 112
Thermal energy, 4
Thermocouple vacuum gauge, 30

Thickness monitor, 85, 86
Torr, 29
Torricelli barometer, 29
Triode sputtering system, 108

V

Vacuum, 30
Vacuum chamber, 46
Vacuum pump, 23
Vacuum pump oil, 61
Vacuum system, 44
Valve, 43, 45
Vapor pressure, 3, 5, 65, 70
Velocity
 evaporated atom, 123
 gas atom, 17, 68
 sound, 17
 sputtered atom, 122
Velocity of sound, 17
Vent, 43
Viton, 46, 52, 54

W

Water molecule, 8
Wehner, G. K., 93, 111, 124